Level 3

3级

美国原版青少年核心能力拓展

好玩的数学

Targeted Mathematics **Student Guided Practice Book**

主 编：〔美〕莎拉·约翰逊

天津出版传媒集团
天津科技翻译出版有限公司

著作权合同登记号：图字：02-2013-306

图书在版编目（CIP）数据

美国原版青少年核心能力拓展：好玩的数学=Targeted mathematics: student guided practice book. 3级：英文/（美）约翰逊（Johnson, S.）主编. —天津：天津科技翻译出版有限公司，2014.4

ISBN 978-7-5433-3344-4

Ⅰ.①美… Ⅱ.①约… Ⅲ.①数学—少儿读物 ②英语—口语—少儿读物 Ⅳ.①O1-49 ②H319.9-49

中国版本图书馆CIP数据核字（2013）第321102号

授权单位：Teacher Created Materials, Inc.
出　　版：天津科技翻译出版有限公司
出 版 人：刘庆
地　　址：天津市南开区白堤路244号
邮政编码：300192
电　　话：022-87894896
传　　真：022-87893237
网　　址：www.tsttpc.com
印　　刷：唐山天意印刷有限责任公司
发　　行：全国新华书店
版本记录：880×1230　16开本　10印张　200千字
2014年4月第1版　2014年4月第1次印刷
定价：25.00元

（如发现印装问题，可与出版社调换）

编者前言

《美国原版青少年核心能力拓展：好玩的数学》(学前阶段及1~8级)是从美国教师创新教材出版公司(Teacher Created Materials Inc.)引进的现行介入式数学指南，全面反映了美国学前班至八年级数学课的现行教学内容及教学要求，同时也为我们展示了美国青少年丰富多彩、生动活泼的学习场景。

阅读使用这套丛书会让你有一种犹如在美国上学、与美国的小朋友同步学习的亲身体验。从中不仅可以了解美国学生在数学课上学些什么，做些什么样的作业，考些什么样的数学题；还可以知道他们的老师在课堂上讲些什么，以及对学生的要求是什么。由此你会发现，他们的数学课与我们的有相同之处，但也并非完全相同。我们侧重于背公式，做习题，备考应试；而他们侧重于理解和掌握数学的基础知识，既讲述初等数学的内容，又介绍了一些高等数学、数论、概率论、统计学的知识，并与其他学科相互联系，从而了解数学在其他学科中的应用，而且在教学中注意联系实际，注重实践应用，因此上数学课不会让学生感到枯燥乏味，而是感觉生动有趣。二者有着不同的教学理念和方式，如果能通过这套丛书的学习将二者有机地结合起来，取长补短，优势互补，必能开阔你的眼界，提高你对数学概念的理解，提升你的应用能力(当然也包括应试能力)。

数学是世界各地通用的一门学科，有着共同的概念、公式、术语、习题、计算方法，因此在这套书中有着非常熟悉的学习内容和知识背景：学过的数学知识，做过的数学习题，考过的数学试题。特别之处在于这套丛书以英文原版形式体现，这就为你营造了一个在熟悉的背景下学习英语的环境，学会用地道的英语来表达学过的知识，表达真实的日常生活和学习活动，学会用英语和同学进行学习互动，从而大幅度提高你的英语水平。既学了数学又学了英语(而且是非常实用的英语)，岂非两全其美的好事。

这套丛书适用于我国广大青少年读者，尤其是双语学校的学生以及打算到英语国家上高中、上大学的学生。学习这套丛书，就等于在国内体验了国外的学校生活，这对今后的深造无疑是大有裨益的。

打开书本，开启你在国内“留学”的全新生活吧！

英语就得天天练——阅读美国孩子的课余英文原版书

好玩的数学——体验美国青少年数学学习的乐趣

Table of Contents

目 录

Table of Contents

目 录

Student Welcome Letter
欢迎信

Dear Student,

You are starting a math program that will help you review third grade. Up to this point in school, you have learned many mathematical skills. This program will help you focus on what you already know how to do in math and what you need to learn. You will learn the important mathematical concepts, skills, and vocabulary so that you are ready for fourth grade.

Sometimes students have trouble learning math. It can seem confusing. This program will help you practice math every day. You will review many things you learned in third grade. Some of those things include multiplication, fractions, and problem solving.

Please sign the bottom of this letter and keep it as a reminder of the skills you learned in this mathematics program. Have fun!

____

Student Signature（学生签名）

Diagnostic Test
摸底测试

1 Which number means five thousands（千）, no hundreds（百）, three tens, and two ones?

(A) 5,320

(B) 5,032

(C) 50,302

(D) 5,203

2 Which group shows the numbers in order from least to greatest?

(F) 7,046 7,024 7,035 7,082

(G) 7,035 7,046 7,082 7,024

(H) 7,082 7,035 7,046 7,024

(J) 7,024 7,035 7,046 7,082

3 25 + 53 can also be written as

(A) (25 + 53) + (25 + 53)

(B) (5 + 5) + (2 + 3)

(C) (20 + 50) + (5 + 3)

(D) (50 + 2) + (30 + 5)

4 Which shows one way to check（检查）the answer?

188 + 51 = 239

(F) 188 – 51

(G) 188 + 239

(H) 51 + 239

(J) 239 – 188

5 Round（取整）384 to the nearest 10.

(A) 380

(B) 390

(C) 374

(D) 274

6 Brian wants to buy favors（聚会的礼品）to give at his party. Each favor costs 10 cents（美分）. Brian has 87 cents. What is the greatest number of favors he can buy?

(F) 9

(G) 8

(H) 7

(J) 6

Diagnostic Test *(cont.)*

7 Which number pair has a product（乘积） of 12?

Ⓐ 3, 4

Ⓑ 9, 3

Ⓒ 24, 2

Ⓓ 17, 5

8 Four friends want to buy a video game（电子游戏）. They each chip in（出钱，集资） 8 dollars. Which of the following could be used to find out how much money the friends have all together?

Ⓕ 4 + 8

Ⓖ 8 ÷ 4

Ⓗ 8 – 4

Ⓙ 4 x 8

9 Five brothers and sisters are sharing（共享，收入分成） $24.00. Each is to receive an equal（相等的） amount. If any money is left over, it will go into a home bank. Which statement（叙述） about the money is true?

Ⓐ $34.00 will go into the home bank.

Ⓑ Each child will receive $4.80.

Ⓒ Each child will receive $13.00 and $9.00 will go into the home bank.

Ⓓ $26.00 will go into the home bank.

10 Omar picked（采摘） 18 oranges from the tree in his backyard（后院，庭院）. He wants to keep half and give half to his grandma. How many oranges would each of them get?

Ⓕ 7

Ⓖ 36

Ⓗ 20

Ⓙ 9

11 Look at the shaded part（阴影区域） of each figure（图）. Which figure is $\frac{2}{3}$ shaded?

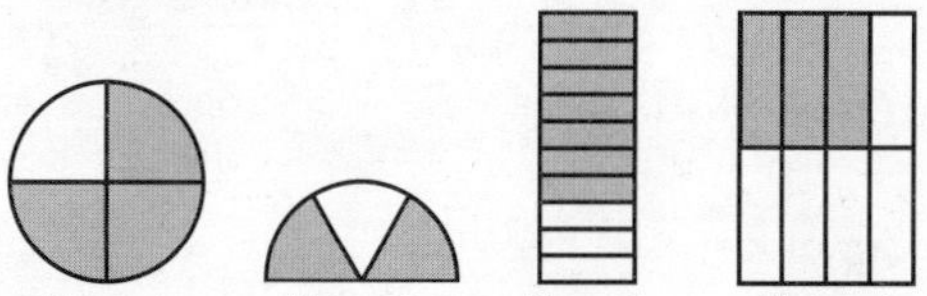

Figure 1 Figure 2 Figure 3 Figure 4

Ⓐ Figure 1

Ⓑ Figure 2

Ⓒ Figure 3

Ⓓ Figure 4

12 Which fraction（分数） stands for（代表） the number of shaded circles（图形）?

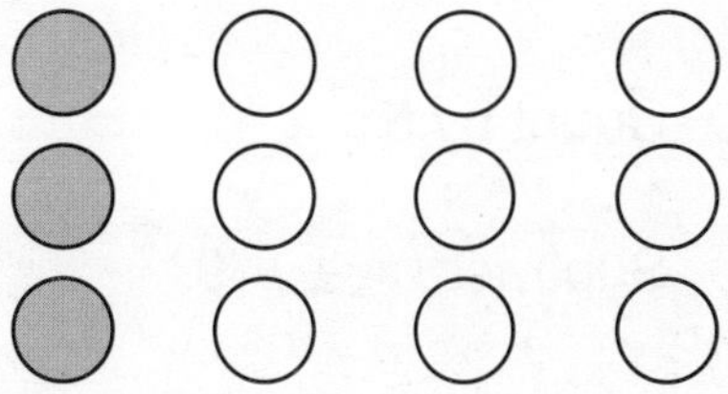

Ⓕ $\frac{1}{3}$ Ⓗ $\frac{1}{2}$

Ⓖ $\frac{1}{4}$ Ⓙ $\frac{3}{4}$

Diagnostic Test *(cont.)*

13 Samantha ate $\frac{2}{8}$ of an apple. What fraction is equivalent to（相当于，等于）$\frac{2}{8}$?

Ⓐ $\frac{1}{2}$

Ⓑ $\frac{1}{5}$

Ⓒ $\frac{1}{4}$

Ⓓ $\frac{1}{3}$

14 Which fraction is less than（大于）$\frac{1}{2}$?

Ⓕ $\frac{4}{5}$

Ⓖ $\frac{3}{4}$

Ⓗ $\frac{2}{3}$

Ⓙ $\frac{1}{3}$

15 What pattern（规律）was used to make this number group?

50, 59, 68, 77

Ⓐ Count by 5's.

Ⓑ Subtract（减去）9.

Ⓒ Multiply by（乘以）2.

Ⓓ Add 10 and subtract 1.

16 A foot（英尺）is a measure of ____________.

Ⓕ temperature（温度）

Ⓖ time

Ⓗ length

Ⓙ volume（容量）

17 Which is an example of perimeter（周长）?

Ⓐ the distance around a playground

Ⓑ the depth of a lake

Ⓒ the height of a building

Ⓓ the length of a straight line（直线）across a field

18 What unit would be best to state the measurement（计量）of a small carton（盒）of milk?

Ⓕ kilogram（千克）

Ⓖ pint（品脱）

Ⓗ inches（英寸）

Ⓙ gram（克）

Diagnostic Test *(cont.)*

19 About how many centimeters（厘米）wide is the starfish（海星）from x to y?

Ⓐ 1 cm　Ⓑ 45 cm　Ⓒ 6 cm　Ⓓ 14 cm

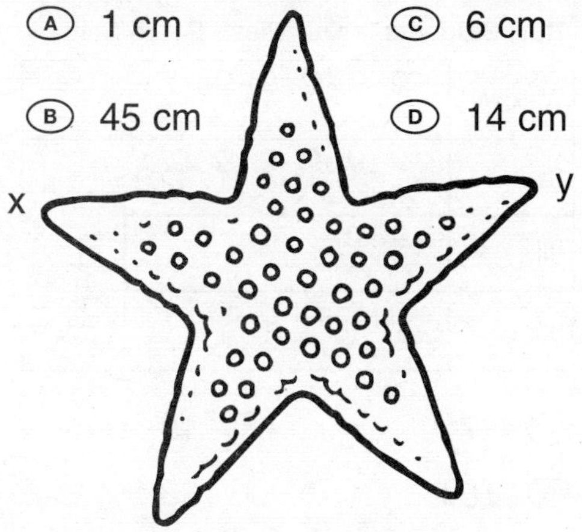

20 Your soccer game（足球赛）began at 4:00 P.M. and ended at 7:00 P.M. How long was the game?

Ⓕ 3 hours long

Ⓖ 1 hour and 30 minutes long

Ⓗ 10 hours long

Ⓙ 2 hours and 30 minutes long

21 How many sides（边）and corners（角）does this shape（形状，图形）have?

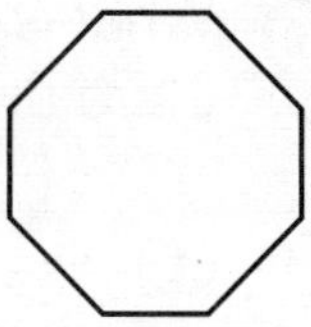

Ⓐ 8 sides and 8 corners

Ⓑ 6 sides and 8 corners

Ⓒ 8 sides and 10 corners

Ⓓ 4 sides and 4 corners

22 Which shape is congruent（一致的）to the rectangle（长方形）below?

Ⓕ Ⓖ Ⓗ Ⓙ

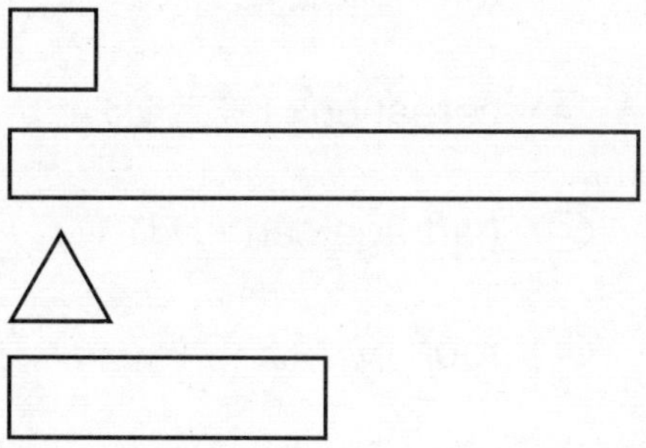

23 Which figure shows a line of symmetry（对称）correctly drawn?

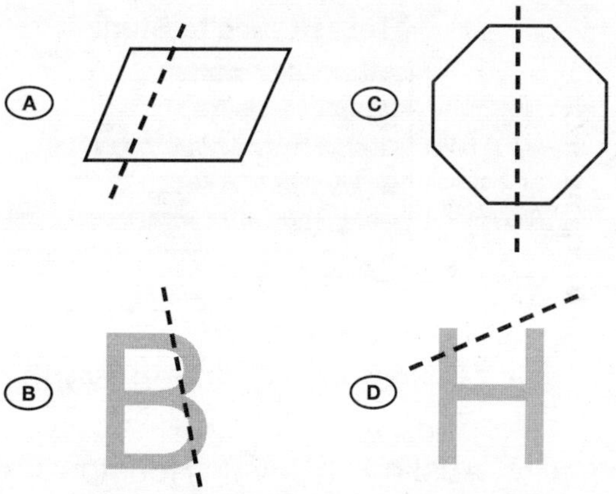

24 Look at the pattern（图案，模式）. Which shape belongs on the line?

▽ ▷ ○ △ ▽ ▷ ○ ___ ▽ ▷

Ⓕ ▷　Ⓗ ▽

Ⓖ ○　Ⓙ △

Diagnostic Test *(cont.)*

25 The shape of a pyramid（金字塔）could be used to create a __________?

Ⓐ television（电视）

Ⓑ cereal box（麦片盒）

Ⓒ garbage can（垃圾箱）

Ⓓ roof（屋顶）

26 Which city received two votes（选票）less than San Antonio?

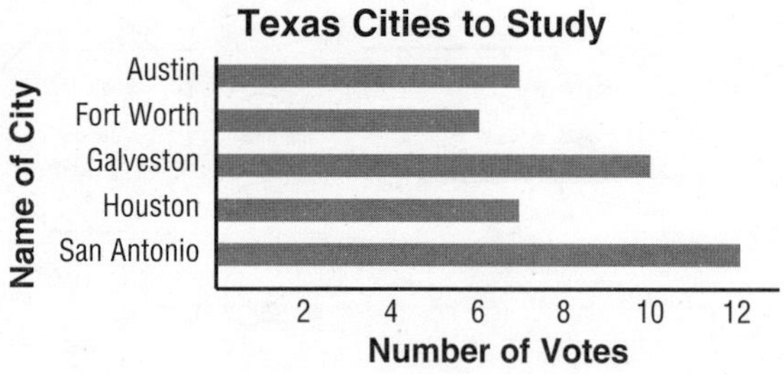

Ⓕ Galveston　Ⓗ Fort Worth

Ⓖ Austin　Ⓙ Houston

27 How many more shirts（衬衫）than sweaters（毛衣）does Cho have?

Ⓐ 12　Ⓒ 6

Ⓑ 10　Ⓓ 2

28 What is the total number（总数）of students in the class?

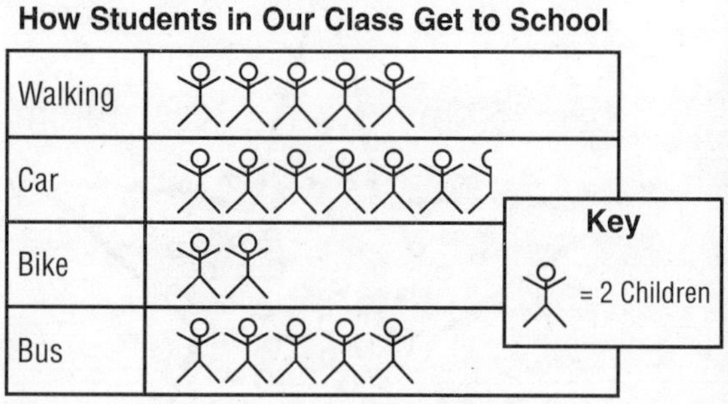

Ⓕ 37　Ⓗ 19

Ⓖ 18　Ⓙ 38

29 What is the best title（最恰当的标题）for the bar graph（条形图）?

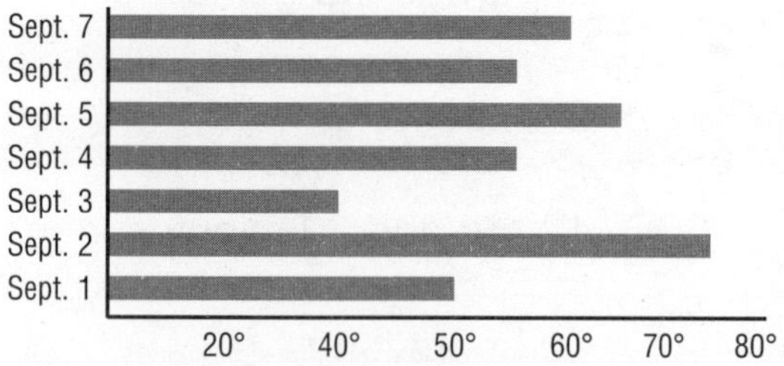

Ⓐ The Number of Days in September
Ⓑ Cool Days During September
Ⓒ How Many Weeks in September
Ⓓ Temperatures During the First Week of September

30 What is the probability（概率）that there are 31 days in August?

Ⓕ likely

Ⓖ impossible

Ⓗ unlikely

Ⓙ certain

Windows

数字窗

Directions: Play this game with a partner（同伴）. You will need number cards and some counters（计数器）in two different colors (one color for each player). Choose four number cards and then arrange（组合）them to make the highest four-digit number（最大的四位数）possible. Say this number aloud. Choose the value of any one digit（任意一位数代表的数值）from this number and cover the corresponding（对应的）window below. For example, if your number is 4,321, you can choose to cover 1, 20, 300, or 4,000. The winner is the first person to cover four windows in a row.

30	6,000	400	4,000
500	5,000	100	2
60	5	40	2,000
1,000	50	4	10
6	300	1	3
200	20	3,000	600

Creating a Table 1

制作表格 1

When a problem has a lot of information（信息）, placing the information in a table helps you organize the information so that it can be easily understood. Using a table can help you keep track of（记录）the information and reduce the chance of making mistakes（减少出错的机会）. A table makes it easy to see what information is given and what information is missing.

When a table is created, the information often shows a pattern（规律）, which can help you solve the problem.

Planning Your Table（规划表格）

- First, read the problem and decide what information is given and what information you need to find out.
- Next, be sure you understand what the table is going to tell you. Include headings（表头）to label the information that you are organizing（标志出你汇总的信息）.

Problem: What's My Age?

The Problem（问题）

Mrs. Stone is 35 years old. Her son, Ty, is 7 years old. How old will Ty be when his mom is 39?

Understanding the Problem（理解问题）

- *What do we know?*
  Mrs. Stone is 35 now.
  Ty is 7 now.
  They will both get one year older every year.
- *What do we need to find out?*
  How old will Ty be when Mrs. Stone is 39?

Planning and Communicating a Solution（规划和交流解决方案）

To solve（解决）this problem, create a table with two columns（列）. Write the heading *Mrs. Stone* above one column and *Ty* above the other column. Write Mrs. Stone's age from 35 to 39 in her column (from top to bottom)（从上至下）. Then, write Ty's ages in the other column.

Hint（提示）**:** Start with 7 because we already know that Ty is 7 when Mrs. Stone is 35.

Mrs. Stone	Ty
35	7
36	8
37	9
38	10
39	11

- *Do you see the answer?*
  When Mrs. Stone is 39 years old, Ty will be 11 years old.

Reflecting and Generalizing（反思与归纳）

The table makes it easy to see a pattern. If you see the pattern right away, you can find the answer easily. For this table, there is a subtraction（减法）pattern. Mrs. Stone's age minus（减去）Ty's age always equals（等于）28 years.

Extension（扩展练习）

How old will Mrs. Stone be when Ty is 20 years old?

Standardized Test Preparation 1
标准考前测试 1

1 Which number means eight thousands, two hundreds, six tens, and five ones?

Ⓐ 8,265

Ⓑ 82,605

Ⓒ 8,605

Ⓓ 8,256

2 What is the correct number to replace（替换） the X on the number line?

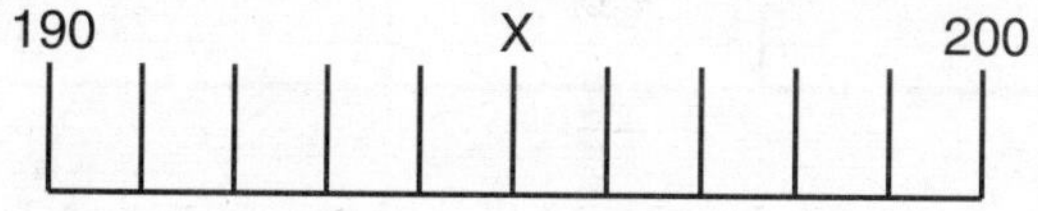

Ⓕ 202

Ⓖ 200

Ⓗ 195

Ⓙ 199

3 179 + ________ = 200

Ⓐ 201

Ⓑ 121

Ⓒ 21

Ⓓ 11

4 Maria and Che are collecting stamps（集邮）. Che has 227 stamps. Maria has 33 fewer（更少的） stamps. How many stamps does Maria have?

Ⓕ 194

Ⓖ 184

Ⓗ 160

Ⓙ 97

5 Round（取整） 868 to the nearest 100.

Ⓐ 870

Ⓑ 800

Ⓒ 900

Ⓓ 1,068

6 Pick one question from this test. Explain（解释） how and why you chose your answer.

Lesson 2

Number Lines
数字直线

Directions: Write the missing（缺失的） numbers on each of the number lines below. Then, make a list（列表，清单） of four-digit numbers（四位数） using the numbers 2, 3, 4, and 8. Write these numbers on the appropriate（适当的） number lines.

1. 2,340 2,350

2. 3,200 3,300

3. 4,000 4,600

4. 3,820 3,850

5. 8,300 9,200

6. 8,200 8,250

7. 2,500 3,000

8. 4,750 4,840

Pet Contest
宠物竞赛

Pet Contest *(cont.)*

What Is the Problem?

A local pet store（当地的宠物店）held a pet contest. There were a number of different categories（奖项类别）, including cutest（最可爱的）pet, fastest（跑得最快的）pet, and loudest（叫声最洪亮的）pet. Your job is to help award the prizes（颁奖）in the category of fastest pet.

Use the information in the problems below to create a table and answer the questions. Before you begin solving each problem, think about what information is given and what information you will need to figure out to complete the problem. Think about how you will create your table and what you will include in（把……列入）it.

Remember: If you have trouble（麻烦）planning ways to solve a problem, reread（重新读）it and rethink（重新思考）your strategy（策略）. When you have a solution（解决办法）, think about whether your answer makes sense（有意义）.

Problem-Solving Strategy（解题策略）: Creating a Table

Problem A

In the race（比赛）, the horse traveled（行进）40 mph*（每小时英里数）. The cat traveled 25 mph. The turtle（乌龟）traveled 1 mph. The dog traveled 30 mph. The pig traveled 8 mph. The mouse（老鼠）traveled 6 mph. Which animals should be awarded the first, second, and third prize?

Hint: Create a table with two columns. Do not forget the heading labels（表头标记）. Place each animal in order（按顺序）, top to bottom, according to its speed（按照速度）.

Problem B

The pet store is offering cash prizes（现金奖励）. The sixth place winner will win（赢得）$10. The fifth place winner will win $20. Each faster animal will receive $10 more than the slower animal before them.

How much will the prize be for first place? Which animal will win that prize?

Hint: Add a third column to the table you created for Problem A in order to figure out the cash prize amounts（数量）.

Problem C

How much faster is the fastest pet than the slowest pet?

Hint: Use the table you have already created. Subtract the speed of the slowest animal from the speed of the fastest animal.

Extension

A cheetah（猎豹）can travel over 50 mph. If a cheetah had been in the race, how would this have changed（改变）the results（结果）?

If there had been seven animals in the race and the 7th place winner won $10, how much would the prize be for the first place winner?

*为体现原版书的特色，书中出现的计量单位在不影响解题思路的情况下均保持原貌，个别在我国不常用的计量单位已做换算标注。计量单位换算可参见附录B：数学用表。

Standardized Test Preparation 2
标准考前测试 2

1 What is the value（数值） of 4 in 14,068?

Ⓐ 4 ones

Ⓑ 4 tens

Ⓒ 4 hundreds

Ⓓ 4 thousands

2 Which group shows the numbers in order from least to greatest（由小到大）?

Ⓕ 88, 72, 64, 61, 50

Ⓖ 44, 52, 66, 35, 72

Ⓗ 76, 35, 22, 9, 18

Ⓙ 15, 24, 41, 49, 56

3 31 + ______ = 40

Ⓐ 8

Ⓑ 9

Ⓒ 19

Ⓓ 29

4 How many flowers belong（属于, 应归入） in the last box?

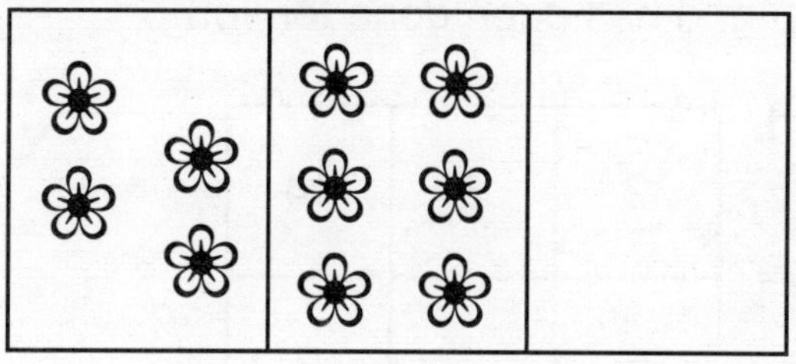

Ⓕ 10　　Ⓗ 7

Ⓖ 8　　Ⓙ 2

5 509 + ______ = 600

Ⓐ 81

Ⓑ 85

Ⓒ 91

Ⓓ 191

6 Pick one question from this test. Explain how and why you chose your answer.

____

____

____

____

____

____

Lesson 3

Make the Number
填 数

Directions: Follow the directions below（根据下面的指示）, using the numbers from the previous grid（前面的网格）, to help you fill in the numbers（填数）. At least the first answer in each grid has been done for you.

1.

9	3	14
5	7	2
12	15	8

2. What would you add to each number in grid 1 to make 18?

9	15	4

3. What would you add to each number in grid 2 to make 100?

91	85	

4. What would you subtract（减） from each number in grid 3 to make 20?

71		

5. What would you subtract from each number in grid 4 to make 50?

21		

6. What would you subtract from each number in grid 5 to make the number you started with in grid 1?

12		

Addition Problems
加法练习题

Directions: Use any addition method you have learned to answer the problems below. Be sure to show your work.

Part A

1. 19 + 23 = ____________________
2. 28 + 44 = ____________________
3. 26 + 37 = ____________________
4. 56 + 41 = ____________________
5. 38 + 11 = ____________________
6. 39 + 85 = ____________________
7. 47 + 68 = ____________________
8. 98 + 19 = ____________________

Part B

9. 67 + 591 = ____________________
10. 48 + 126 = ____________________
11. 457 + 39 = ____________________
12. 206 + 26 = ____________________
13. 368 + 93 = ____________________
14. 231 + 56 = ____________________
15. 166 + 78 = ____________________
16. 19 + 908 = ____________________

Part C

17. $\begin{array}{r} 455 \\ +\ 779 \\ \hline \end{array}$

18. $\begin{array}{r} 396 \\ +\ 222 \\ \hline \end{array}$

19. $\begin{array}{r} 688 \\ +\ 261 \\ \hline \end{array}$

20. $\begin{array}{r} 421 \\ +\ 493 \\ \hline \end{array}$

21. $\begin{array}{r} 287 \\ +\ 333 \\ \hline \end{array}$

22. $\begin{array}{r} 685 \\ +\ 234 \\ \hline \end{array}$

23. $\begin{array}{r} 259 \\ +\ 788 \\ \hline \end{array}$

24. $\begin{array}{r} 884 \\ +\ 117 \\ \hline \end{array}$

Creating a Table 1 Problems
制作表格 1 练习

Directions: Solve each problem and show your work.

Problem A

Janine earns（赚得） 10¢（¢, 美分） a day. She saves（省下, 存下） half of her money each day. How long will it take Janine to save 50¢? What is the total amount（总数） of money earned by Janine?

Day								
Money Earned								
Money Saved								

Problem B

Ling is picking（采摘） apples. Every three minutes（分钟） she puts five apples in the basket（篮子）. How long before Ling has 30 apples in the basket?

Minutes	**Apples in Basket**

Standardized Test Preparation 3
标准考前测试 3

1 Yo-Yos（悠悠球） cost $2.75. Toy bears（玩具熊） cost twice as much. Sahl bought（买） 2 toy bears. How much did Sahl spend（花费）?

Ⓐ $33.00

Ⓑ $21.75

Ⓒ $11.00

Ⓓ $7.25

2 How many dots（圆点） should be in the empty circle（空白的圆圈）?

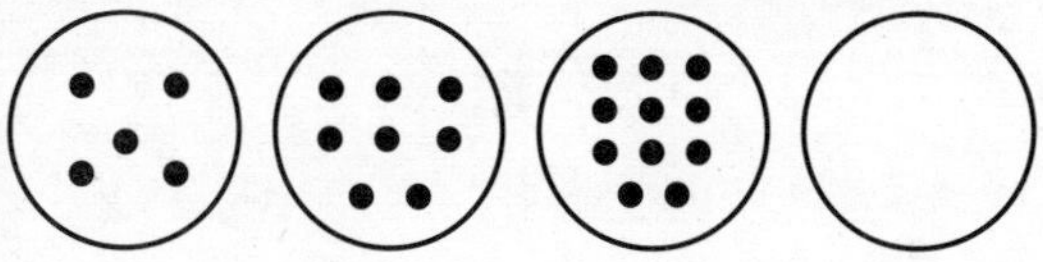

Ⓕ 14

Ⓖ 12

Ⓗ 10

Ⓙ 8

3 What is the value of 3 in 18,372?

Ⓐ 3 thousands

Ⓑ 3 hundreds

Ⓒ 3 tens

Ⓓ 3 ones

4 What is the next number in the sequence（序列）?

28, 33, 38, 43, ____

Ⓕ 44

Ⓖ 48

Ⓗ 50

Ⓙ 53

5 There were 105 boxes of chalk（粉笔） in the supply room（库房）. Teachers took 76 boxes of chalk. How many boxes of chalk were left（剩下的）?

Ⓐ 39 boxes of chalk

Ⓑ 31 boxes of chalk

Ⓒ 29 boxes of chalk

Ⓓ 25 boxes of chalk

6 Pick one question from this test. Explain how and why you chose your answer.

Number Sacks
数字袋

Directions: Make four number sentences（算式） in each box below by choosing one number from sacks 1, 2, and 3 and two signs（运算符号） from sack 4. Use your knowledge of inverse relationships（相反关系） to help you.

Sack 1 **Sack 2** **Sack 3** **Sack 4**

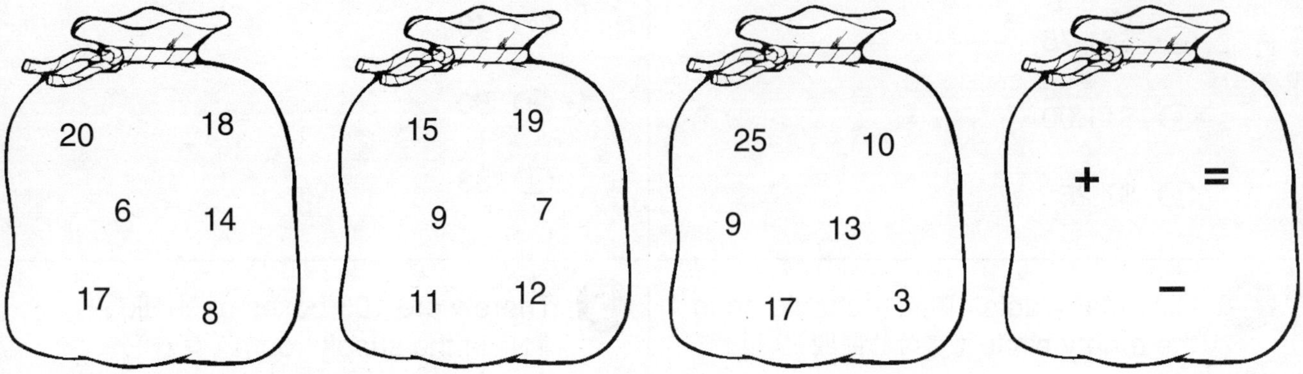

Example: **6 + 11 = 17** **17 − 11 = 6** **17 − 6 = 11** **11 + 6 = 17**		

Subtraction Problems
减法练习

Directions: Use any subtraction method you have learned to answer the following problems. Be sure to show your work.

Part A

1. 56 – 37 = ____________

2. 46 – 19 = ____________

3. 29 – 8 = ____________

4. 80 – 16 = ____________

5. 60 – 17 = ____________

6. 37 – 18 = ____________

7. 53 – 16 = ____________

8. 92 – 28 = ____________

Part B

9. 221 – 93 = ____________

10. 451 – 87 = ____________

11. 363 – 45 = ____________

12. 115 – 78 = ____________

13. 182 – 45 = ____________

14. 302 – 96 = ____________

15. 911 – 28 = ____________

16. 708 – 39 = ____________

Part C

17. $\begin{array}{r} 461 \\ -296 \\ \hline \end{array}$

18. $\begin{array}{r} 502 \\ -187 \\ \hline \end{array}$

19. $\begin{array}{r} 615 \\ -149 \\ \hline \end{array}$

20. $\begin{array}{r} 236 \\ -168 \\ \hline \end{array}$

21. $\begin{array}{r} 522 \\ -479 \\ \hline \end{array}$

22. $\begin{array}{r} 506 \\ -129 \\ \hline \end{array}$

23. $\begin{array}{r} 927 \\ -686 \\ \hline \end{array}$

24. $\begin{array}{r} 801 \\ -238 \\ \hline \end{array}$

Buying School Supplies
购买学习用品

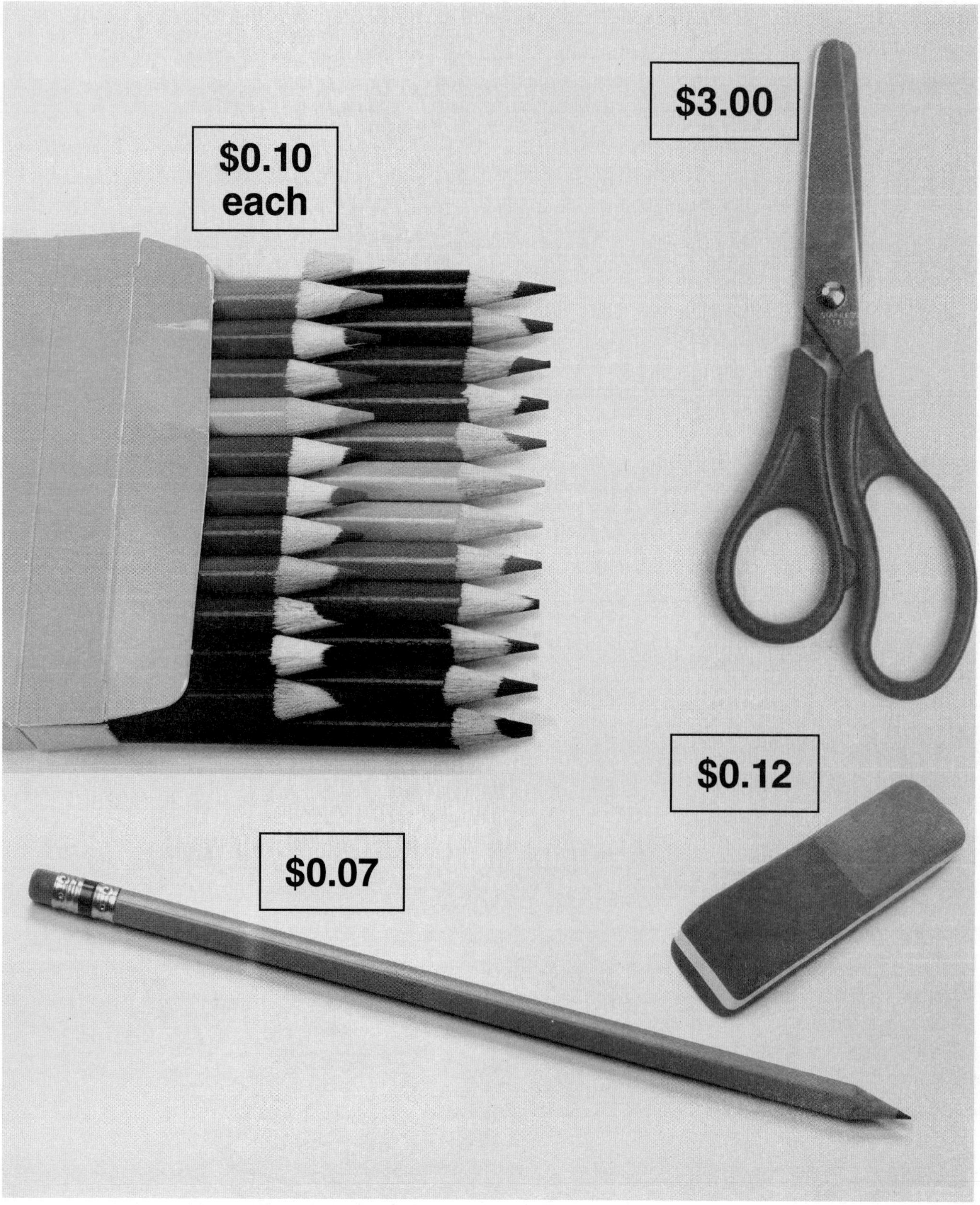

Buying School Supplies *(cont.)*

What Is the Problem?

Your school has a store that sells school supplies to students. There are many items（物品） for sale. Use the information from the picture and the problems below to answer the questions. You will need to create a table to solve the problems.

Before you begin solving each problem, think about what information is given and what information you will need to figure out to complete the problem. Think about how you will create your table and what you will include in it.

Remember: If you have trouble planning ways to solve a problem, reread it and rethink your strategy. When you have a solution, think about whether your answer makes sense.

Problem-Solving Strategy: Creating a Table

Problem A

A pencil costs \$0.07. How much will nine pencils cost?

Hint: In your table, one column will have the number of pencils and one column will have the cost of the pencils.

Problem B

You want to buy a pencil for everyone in your class. There are 20 students in your class. How much money will you need?

Hint: Continue（继续） adding columns to your table until you have 20 total pencils in the first column.

Problem C

Ted has \$0.33. If he buys 3 pencils, how many erasers（橡皮） can he buy with his leftover（剩余的） money?

Hint: Look at your table above to calculate（计算）how much 3 pencils will cost. Then, subtract that amount from \$0.33 to figure out（计算出） how much money Ted has left.

Extension

Colored pencils cost \$0.03 more than the regular（普通的） pencils. How much money would you need to buy 5 colored pencils?

Hint: One column of your table will show the costs of regular pencils and the second column will show the costs of the colored pencils.

Standardized Test Preparation 4

标准考前测试 4

1 Koko's grandmother lives 263 miles（英里） away. Koko and her family have driven 67 miles. How many more miles do they need to drive?

Ⓐ 295

Ⓑ 213

Ⓒ 196

Ⓓ 190

2 Tony has $68.00. He needs $300.00 to buy supplies（物资） to repair his tree house. How much more does he need before he can buy the supplies?

Ⓕ $212.00

Ⓖ $119.00

Ⓗ $232.00

Ⓙ $242.00

3 349 – 87 = ______

Ⓐ 263

Ⓑ 262

Ⓒ 249

Ⓓ 248

4 Which number means five thousands, no hundreds, three tens, and two ones?

Ⓕ 5,320

Ⓖ 5,032

Ⓗ 50,302

Ⓙ 5,203

5 How many stars should appear（出现） in the empty（空的） box?

Ⓐ 4

Ⓑ 13

Ⓒ 10

Ⓓ 11

6 Pick one question from this test. Explain how and why you chose your answer.

____

____

____

____

____

____

Number Line 0–1,000

数字直线0-1000

Directions: Record（记录） these numbers on page 24 Part A on the correct places on the number line below.

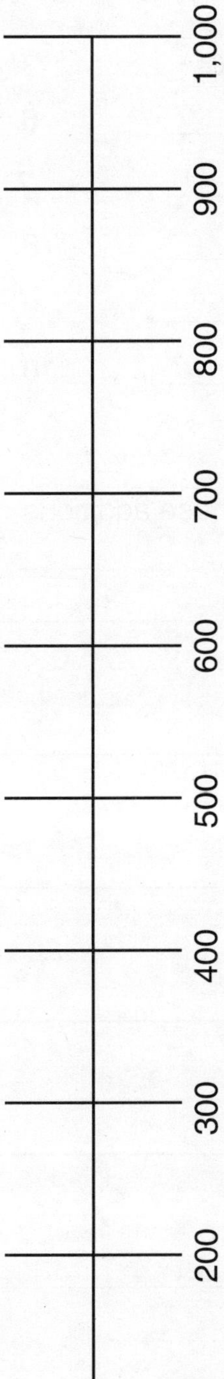

Rounding
四舍五入

Directions: Round each group（组）of numbers below as instructed（指导）.

Part A

Round these numbers to the nearest 100:

1. 68 __________

2. 142 __________

3. 607 __________

4. 493 __________

5. 586 __________

6. 545 __________

7. 750 __________

8. 39 __________

9. 221 __________

10. 971 __________

Part B

Estimate（估算）the results（结果）of these additions（加法）and subtractions（减法）by rounding the numbers to the nearest 10:

11. 47 + 35 50 + 40 = 90

12. 123 – 49 __________

13. 61 – 22 __________

14. 361 + 412 __________

15. 63 + 91 __________

16. 449 + 568 __________

17. 903 – 198 __________

18. 225 + 177 __________

Creating a Table 1 Group Problems
制作表格 1 分组练习

Directions: Solve Problem A and show your work. Using the strategy you've studied this week, write and solve your own（自己的）problems for Problem B and Problem C.

Problem A

Every day the Wilsons put money into the family's piggy bank（存钱罐）. Dad puts in 10 cents（美分）. Mom puts in 15 cents. Junior puts in 5 cents. How much money is in the piggy bank at the end of one week? How much money does each family member put in the piggy bank?

Problem B: ____________________

____

____

____

____

____

Show work:

Answer: ____________________

Problem C: ____________________

____

____

____

____

____

Show work:

Answer: ____________________

Standardized Test Preparation 5
标准考前测试 5

1 Which problem has an answer that is a one-digit number（一位数）?

Ⓐ 59 – 1

Ⓑ 88 – 22

Ⓒ 4 + 5

Ⓓ 9 + 9

2 Round 384 to the nearest 10.

Ⓕ 380

Ⓖ 390

Ⓗ 374

Ⓙ 400

3 (6 + 2) + (7 + 8) = _____

Ⓐ (6 + 7) – (2 + 8)

Ⓑ (8 – 6) + (2 + 7)

Ⓒ 8 + 15

Ⓓ 15 – 8

4 Round 346 to the nearest 10.

Ⓕ 350

Ⓖ 345

Ⓗ 300

Ⓙ 340

5 Which would best be measured in hours（以小时计算）?

Ⓐ cooking an egg

Ⓑ driving across Texas（开车穿越得克萨斯州）

Ⓒ putting on your coat

Ⓓ reading a poem（诗）

6 Pick one question from this test. Explain how and why you chose your answer.

____

____

____

____

____

____

Lesson 6

Don’t Count on It! 1

学乘法不数数 1

Directions: Use multiplication（乘法）or addition to find the total number of stars in each of these patterns（图案，模式）. In the box next to each pattern, write a number sentence to show how you found the answer. Then, solve your number sentence. DO NOT COUNT!

Example:

4 rows（行）x 5 columns（列）

4 x 5 = 20

1.

2.

3.

Don't Count on It! 1 *(cont.)*

4.

5.

6.

Don't Count on It! 2

学乘法不数数 2

Directions: Draw an array（阵列）to solve each problem. Then, write the answers in the boxes.

1. 3 x 4 = ☐ **2.** 7 x 6 = ☐ **3.** 4 x 4 = ☐

4. 8 x 9 = ☐ **5.** 5 x 7 = ☐ **6.** 9 x 2 = ☐

7. 7 x 3 = ☐ **8.** 6 x 1 = ☐ **9.** 4 x 6 = ☐

Guessing and Checking
推测和验算

Guessing and checking can be used to solve a variety of problems. Begin by looking for the important facts in the problem and determining（决定）what you need to solve. Then make an educated guess（有根据的推测）. Base the guess on information you already know or ideas that make sense. Don't make a wild guess（胡乱猜想）. Then, check your answer. If your educated guess is wrong, decide if it is too small or too large, and adjust（调整）your guess. Check your answer again. Repeat these steps until you find the solution.

Learning how to make reasonable（合理的）guesses takes practice（练习）. Creating a table will help you keep track of your guesses and results (checks) until you find the correct（正确的）solution.

Problem: Adding Ages

The Problem

Alan is five years older than Juan. Alan's age plus Juan's age equals 25. What are their ages?

Understanding the Problem

- *What do we know?*
  Alan is five years older than Juan.
  Alan's age plus Juan's age equals 25.
- *What do we need to find out?*
  What are Alan's and Juan's ages?

Planning and Communicating a Solution

To make an educated guess, you need a starting point（起始点）. Let's make a first guess that Alan is 12. That means that Juan is 7, because he is five years younger than Alan.

Place this information in a table and check the guess by adding the ages.

Alan's Age	Juan's Age	Total
12	7	19

You can see from your table that the first guess is too low. You must rethink（重新考虑）your guess to find a larger total. Add a row to the table so you can make a new higher guess.

Let's guess that Alan is 18.

Alan's Age	Juan's Age	Total
12	7	19
18	13	31

- *What happened?*
  Now the total is too high. You may have to try another guess or two before you arrive at the solution.

Alan's Age	Juan's Age	Total
12	7	19
18	13	31
15	10	25

- *Do you see the answer?*
  Alan is 15 years old and Juan is 10 years old. 15 plus 10 equals 25.

Reflecting and Generalizing

Because the results have been recorded in the table, you can see exactly what guesses you have made. You will also notice（注意）how closely（接近地）you came to solving the problem each time.

Extension

When Alan is 21, what will be the total of their ages?

Standardized Test Preparation 6
标准考前测试 6

1 9 x 8 = ____

Ⓐ 62

Ⓑ 72

Ⓒ 98

Ⓓ 73

4 Which represents（表达）the same number as 3 x 5?

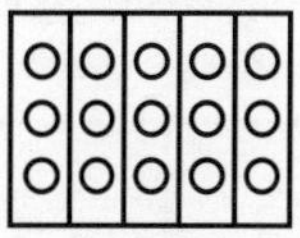

Ⓕ

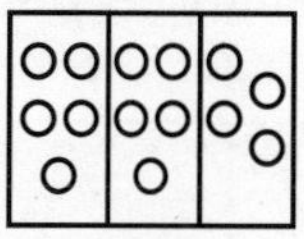

Ⓗ

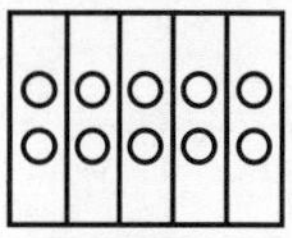

Ⓖ

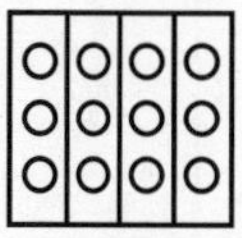

Ⓙ

2 Note the time on the clock. What will the time be in 10 minutes（分钟）?

Ⓕ 4:10

Ⓖ 10:10

Ⓗ 5:00

Ⓙ 4:00

5 What is the name of this shape?

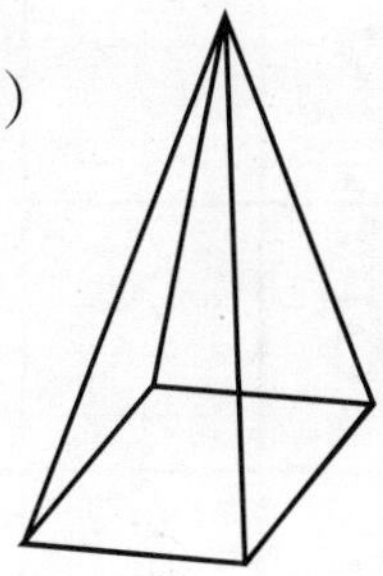

Ⓐ hemisphere（半球）

Ⓑ square pyramid（正四棱柱）

Ⓒ cube（立方体）

Ⓓ triangular prism（三棱柱）

3 Jung's sister grabbed（抓住）a box of crackers（饼干）and ripped（撕）it open. She threw（扔）some of the crackers on the floor. The box showed that it contained（容纳）125 crackers. There were 117 crackers left in the box. How many crackers did Jung's sister throw on the floor?

Ⓐ 20 crackers

Ⓑ 18 crackers

Ⓒ 8 crackers

Ⓓ 5 crackers

6 Pick one question from this test. Explain how and why you chose your answer.

____

____

____

____

____

____

Lesson 7

Multiplication Grid

乘法表

Directions: Fill in all the squares（方格）on the multiplication grid. Start with the numbers that you know. Try to find a pattern to help you solve the harder（更难的）problems.

10										
9										
8										
7										
6										
5										
4										
3										
2										
1										
x	1	2	3	4	5	6	7	8	9	10

On the Farm
农场里的数学

Where does your food come from? Most of it comes from farms. Some farms raise crops（种植作物）such as corn（玉米）or wheat（小麦）, while other farms raise animals. There are also farms that raise both crops and animals. On many farms, you will find cows, horses, pigs, chickens, and more. However, not all farms have animals.

Some farmers raise cows for milk. Mystic Valley Dairy in Wisconsin milks almost 400 cows every day. The dairy farm produces a lot of milk! On another farm in Montana, farmers raised 410,000 chickens. These chickens laid 94 million eggs（下蛋）in one year. That is a lot of eggs!

Farming is a hard job. Many farmers get up before four o'clock in the morning. They spend all day working in their fields and caring for their animals.

COREL

COREL

On the Farm *(cont.)*

What Is the Problem?

As a farmer, you must know how many animals you have. If you are selling food, you need to keep track of how much your farm is producing（生产）. Sometimes it gets confusing（困惑）because the animals are spread out（分散的）in different parts of your farm. They usually like to be together, but sometimes you cannot see them all in one place.

Use the information below to answer the questions. Before you begin solving each problem, be sure to locate（找出）the information you will need to make a strong first guess.

Remember: If you have trouble planning ways to solve a problem, reread it and rethink your strategy. When you have a solution, think about whether your answer makes sense.

Problem-Solving Strategy: Guessing and Checking

Problem A

A farmer has two hen coops（鸡笼）. One is red and one is blue. The farmer collected 405 total eggs yesterday from both coops. He collected 15 more eggs from the hens in the red coop. How many eggs were collected from the hens in the blue coop?

Hint: In your table, first guess how many eggs were collected from the hens in the red coop.

Problem B

Six animals are on the farm. The six animals have a total of 22 legs. The animals are either chickens or cows. How many cows are there? How many chickens?

Hint: Guess how many chickens and cows there are. Then, check your guess to see if it is correct.

Problem C

The farm has 15 milk cows. The average（平均的）cow on the farm produces 10 gallons（加仑）of milk a day. On Monday, the farmer's wife packaged（包装）two times as many（两倍）milk gallons as the farmer. How much milk was packaged by the farmer himself?

Hint: First, multiply to find out how much milk is produced in a day.

Extension

Pretend（假设）that you start your own farm. You want to raise hens for eggs and cows for milk. You have 24 hens on your farm. The hens lay（下蛋）an average of 3 eggs a day. The eggs sell for $2.50 a dozen（一打, 12个）. The cows produce an average of 10 gallons of milk a day, which sell for $3.00 a gallon. If you want to make $315.00 per day, how many cows do you need to have on your farm?

Hint: There are 12 eggs in each dozen sold.

Standardized Test Preparation 7
标准考前测试 7

1 Which number is not a multiple（倍数）of 3?

Ⓐ 23

Ⓑ 21

Ⓒ 30

Ⓓ 24

2 Which of these numbers is a multiple of 9?

Ⓕ 10

Ⓖ 19

Ⓗ 27

Ⓙ 29

3 228 + 692 =

Ⓐ 926

Ⓑ 920

Ⓒ 910

Ⓓ 819

4 At the pet shop, goldfish cost $2.10. Turtles cost twice as much. Sanjeev buys 3 turtles. How much does Sanjeev spend?

Ⓕ $1.26

Ⓖ $6.30

Ⓗ $8.40

Ⓙ $12.60

5 Which of these shapes has more than 7 corners?

Ⓐ pentagon（五边形）

Ⓑ hexagon（六边形）

Ⓒ octagon（八边形）

Ⓓ square（正方形）

6 Pick one question from this test. Explain how and why you chose your answer.

____

____

____

____

____

____

Hundreds Chart
百数表

1	2	3	4	5	6	7	8	9	10
11	12	13	14	15	16	17	18	19	20
21	22	23	24	25	26	27	28	29	30
31	32	33	34	35	36	37	38	39	40
41	42	43	44	45	46	47	48	49	50
51	52	53	54	55	56	57	58	59	60
61	62	63	64	65	66	67	68	69	70
71	72	73	74	75	76	77	78	79	80
81	82	83	84	85	86	87	88	89	90
91	92	93	94	95	96	97	98	99	100

Guessing and Checking Problems
推测和验算 练习

Directions: Solve each problem and show your work.

Problem A
Arrange the numbers 5, 6, 7, 8, 9, and 10 in the shape below so each side of the triangle（三角形）equals 21. Each number may only be used one time.

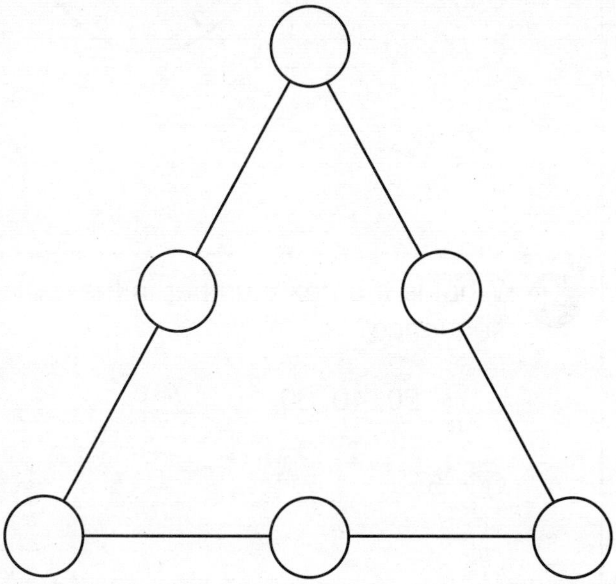

Problem B
Use the numbers 4, 5, 6, 7, 8, 9, 10, 11, and 12 to complete the magic（魔法的）square. All rows and columns must equal 24. Each number may only be used one time.

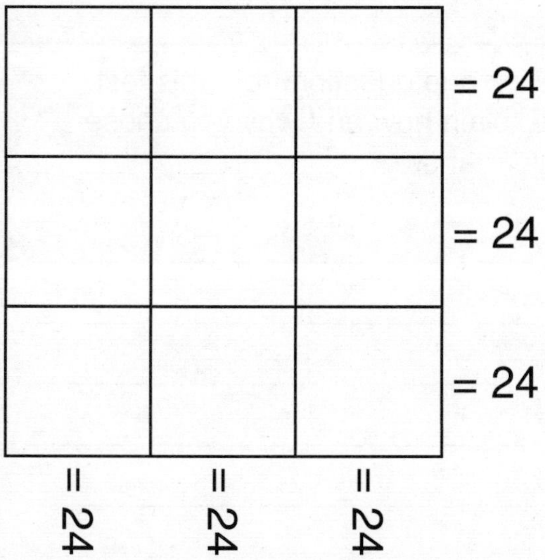

Standardized Test Preparation 8

标准考前测试 8

1 What is the next number sentence in the following sequence（序列）?

1 x 5 = 5

2 x 5 = 10

3 x 5 = 15

Ⓐ 1 x 3 = 3

Ⓑ 3 x 2 = 6

Ⓒ 4 x 5 = 20

Ⓓ 10 x 5 = 15

2 Which represents（代表，表示） the same number as 3 x 4?

Ⓕ

Ⓗ

Ⓖ

Ⓙ

3 Bibi is putting his baseball cards in groups of 5（5组）. All of his cards fit equally（相等的）. Which number could represent how many baseball cards Bibi has?

Ⓐ 12

Ⓑ 55

Ⓒ 46

Ⓓ 72

4 What time is shown on this clock?

Ⓕ 8:40

Ⓖ 8:20

Ⓗ 4:08

Ⓙ 3:40

5 What is the next number in the sequence?

50, 40, 30, 20, ____

Ⓐ 5

Ⓑ 35

Ⓒ 10

Ⓓ 0

6 Pick one question from this test. Explain how and why you chose your answer.

____

____

____

____

____

____

Missing Numbers

算式填空

Directions: Fill in the missing factors（要素）. Use the *Multiplication Grid* on page 32 if you are unsure of your facts.

1. 8 x ____ = 48

2. 6 x ____ = 48

3. 12 x ____ = 48

4. 9 x ____ = 63

5. ____ x 7 = 63

6. 7 x ____ = 63

7. 4 x ____ = 36

8. 9 x ____ = 36

9. ____ x 6 = 36

10. 48 ÷ 6 = ____

11. 48 ÷ 8 = ____

12. 48 ÷ 12 = ____

13. 64 ÷ 8 = ____

14. 81 ÷ 9 = ____

15. 36 ÷ 12 = ____

16. 49 ÷ 7 = ____

17. 35 ÷ 7 = ____

18. 84 ÷ 12 = ____

Make Four
组成四个算式

Directions: Choose three of the numbers below to make four multiplication and division（除法的）equations. The first one has been done for you. You may use the numbers more than once.

24	9	10	20	36	6
18	2	63	42	7	8
21	3	4	30	28	5

Numbers	Multiplication/Division Equations			
1. 9, 7, 63	9 x 7 = 63	7 x 9 = 63	63 ÷ 9 = 7	63 ÷ 7 = 9
2. ______	______	______	______	______
3. ______	______	______	______	______
4. ______	______	______	______	______
5. ______	______	______	______	______
6. ______	______	______	______	______

Challenge

Directions: Choose one of the equations above and rewrite it as a word problem.

____

____

____

____

____

How to Play Football
橄榄球赛里的数学

You are going to a football game. What will you see? Look at the diagram（图表）. It shows a football field. The field is 120 yards（码，1码=3英尺=0.914米）long. Each team has an end zone（球门区）. An end zone is 10 yards at either end of the field. Between the two end zones are 100 yards.

Football is played by two teams. Teams try to get the ball to the end zones. A team can score points（得分）in different ways. The team can complete a pass（传球）or run the ball into an end zone. This is called a touchdown（触地得分）. Touchdowns are worth 6 points. After a touchdown, a team can try to get extra（额外的）points in two ways. A team can get 1 point for kicking the ball through the goalposts（球门柱）or 2 points for completing a pass or running the ball into the end zone. If the team is not going to get a touchdown, it can still try to get points for just kicking a field goal（射门得分）. The ball must go through the goalposts. A field goal earns 3 points.

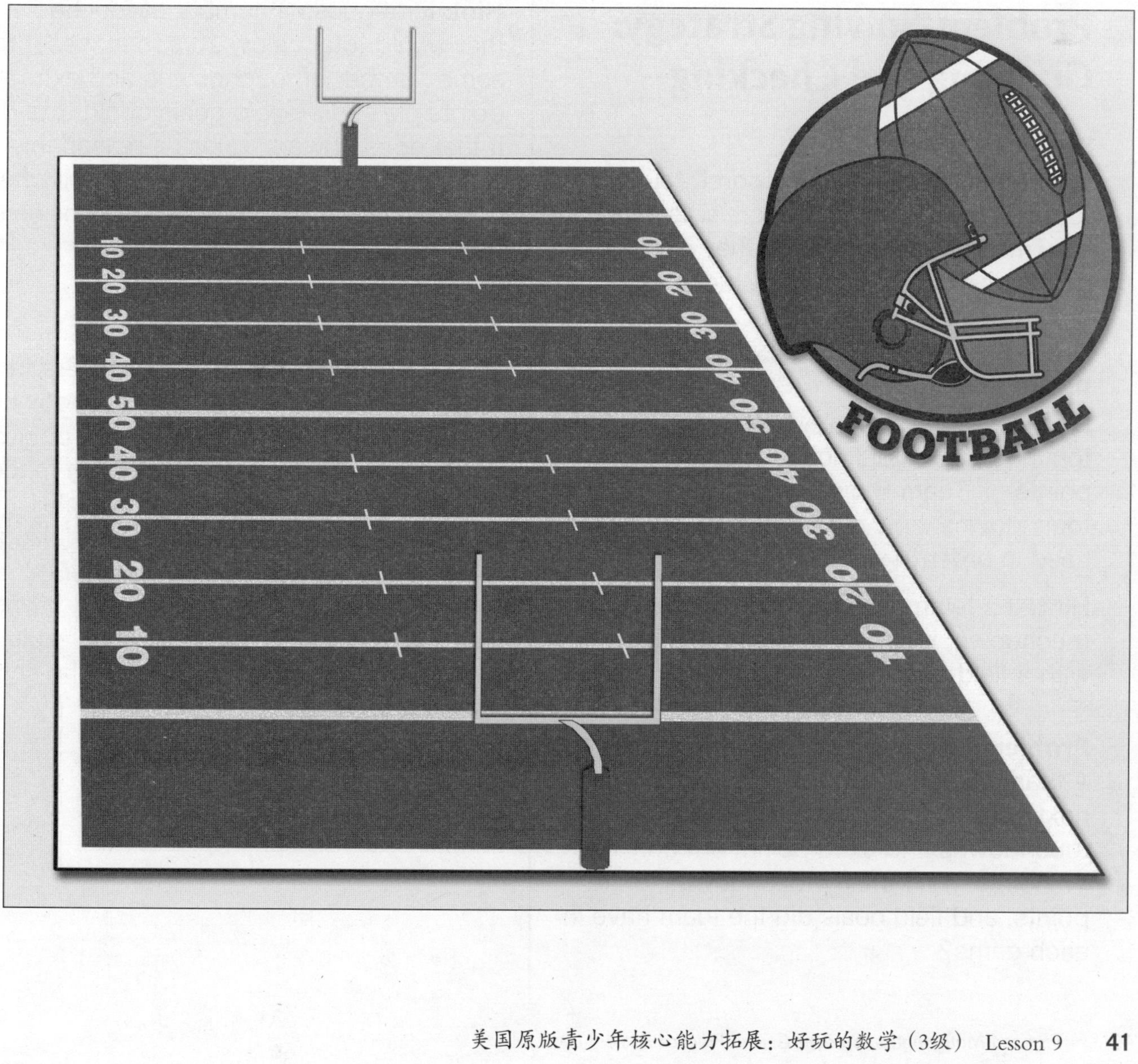

How to Play Football *(cont.)*

What Is the Problem?

Football is an exciting game to watch. Many fans get excited as they watch the two teams compete against（与……竞争）one another and score（得分）touchdowns and field goals. You can help keep track of the scores for the football games in the guessing and checking problems below.

Use the information on page 41 and below to answer the questions. Before you begin solving each problem, be sure to locate the information you will need to make a strong first guess.

Remember: If you have trouble planning ways to solve a problem, reread it and rethink your strategy. When you have a solution, think about whether your answer makes sense.

Problem-Solving Strategy: Guessing and Checking

Problem A

During one game, a team scored four touchdowns. Their final score was 27. For how many touchdowns did they earn extra points and how did they get those extra points?

Problem B

During a game, Team A had two touchdowns. It did not earn any extra points. If Team B did not make any touchdowns, how many field goals would it need in order to win the game?

Hint: If a team does not make a touchdown, it can try to get 3 points with a field goal.

Problem C

A team scored 20 points in a game. In the next game, it scored the same number of touchdowns and extra points but only had 17 points. How many touchdowns, extra points, and field goals did the team have in each game?

Hint: First guess a number of touchdowns that would total below 20. Then, see if the same number of touchdowns and extra points plus some field goals could total 17. If that doesn't work, guess a reasonable（合理的）number (up or down), and check to see if it will work. Check your answers using division and multiplication.

Extension

The scores for Team A and Team B add up to 38. Both teams had three touchdowns, but Team B did not earn any extra points for their touchdowns. Neither team made any field goals. The only extra points earned by Team A were made by kicking the ball through the goal posts after touchdowns. For how many touchdowns was Team A successfully（成功地）able to earn the extra points? What was the score for each team?

Standardized Test Preparation 9

标准考前测试 9

1 8 ÷ 4 = ______

Ⓐ 2

Ⓑ 4

Ⓒ 32

Ⓓ 84

2 7 x ______ = 14

Ⓕ 3

Ⓖ 2

Ⓗ 5

Ⓙ 4

3 Which number means five thousands, seven hundreds, three tens, and two ones?

Ⓐ 5,320

Ⓑ 5,732

Ⓒ 50,372

Ⓓ 5,273

4 How many sides does this hexagon (六边形) have?

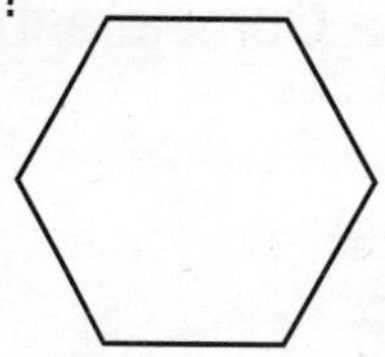

Ⓕ 5

Ⓖ 6

Ⓗ 7

Ⓙ 8

5 Which type of pet do most students have at home?

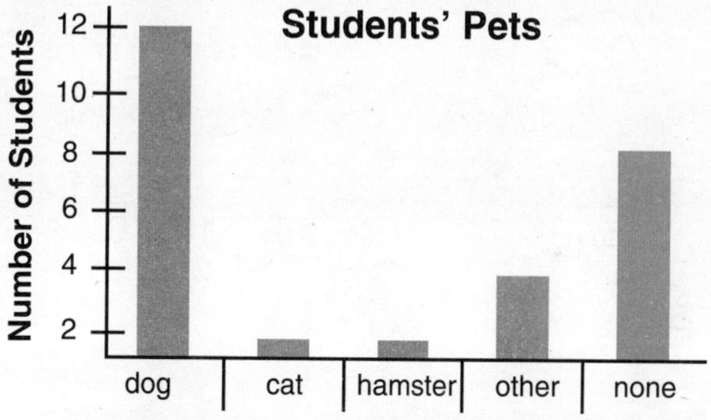

Ⓐ dog

Ⓑ cat

Ⓒ hamster (仓鼠)

Ⓓ other

6 Pick one question from this test. Explain how and why you chose your answer.

____

____

____

____

____

____

Lesson 10

Divide and Check

相除并验算

Directions: Complete these problems. Check your answer using multiplication（乘法）.

1. $1\overline{)10}$
2. $6\overline{)48}$
3. $10\overline{)90}$
4. $4\overline{)36}$
5. $12\overline{)60}$
6. $3\overline{)27}$
7. $8\overline{)56}$
8. $7\overline{)49}$
9. $5\overline{)55}$
10. $10\overline{)40}$
11. $12\overline{)84}$
12. $11\overline{)99}$
13. $4\overline{)24}$
14. $3\overline{)33}$
15. $4\overline{)36}$
16. $6\overline{)36}$
17. $7\overline{)63}$
18. $9\overline{)63}$
19. $9\overline{)108}$
20. $12\overline{)108}$

Guessing and Checking
Group Problems
推测和验算 分组练习

Directions: Solve Problem A and show your work. Using this strategy, write and solve your own problems for Problem B and Problem C.

Problem A

The O'Connor kids' ages total 33 years. Corie is twice（两倍于）as old as Aidan and three times older than Dillon. How old are Corie, Aidan, and Dillon?

Problem B: ______________________________

____

____

____

____

____

Show work:

Answer: ______________________________

Problem C: ______________________________

____

____

____

____

____

Show work:

Answer: ______________________________

Standardized Test Preparation 10

标准考前测试 10

1 8 x 4 = ____

Ⓐ 22

Ⓑ 48

Ⓒ 32

Ⓓ 12

2 What time is shown on this clock?

Ⓕ 3:10

Ⓖ 2:50

Ⓗ 10:15

Ⓙ 10:03

3 How much water is in this jug（水壶）?

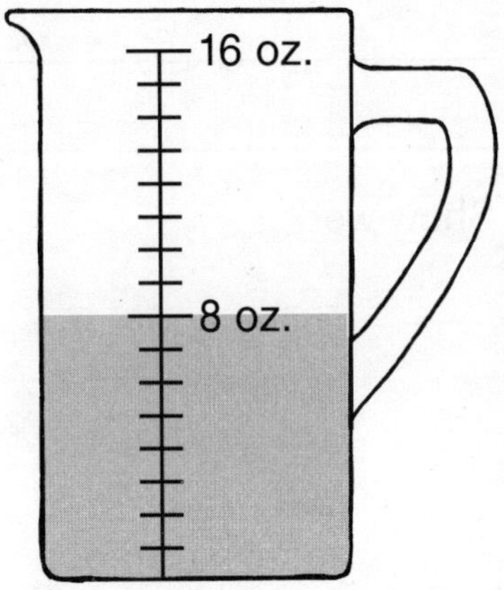

Ⓐ 10 oz.*

Ⓑ 8 oz.

Ⓒ 25 oz.

Ⓓ 16 oz.

* oz.，液体盎司，为容量计量单位。

4 How many students do not have a pet at home?

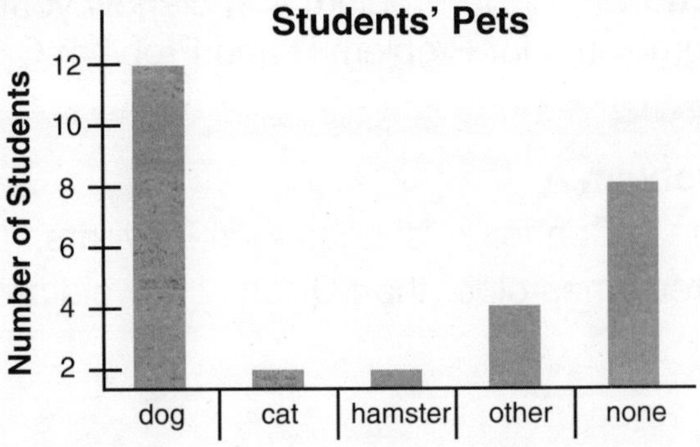

Ⓕ 2

Ⓖ 10

Ⓗ 8

Ⓙ 12

5 What is the missing number in the sequence?

____, 11, 13, 15, 17

Ⓐ 9

Ⓑ 8

Ⓒ 7

Ⓓ 6

6 Pick one question from this test. Explain how and why you chose your answer.

____

____

____

____

____

____

Fractions Fun

有趣的分数

Directions: Choose four fractions. Divide（划分）each rectangle（长方形）into the appropriate sections（适当的区域）, and shade（涂上阴影）each according to the fraction you chose. Complete the addition（加法）problem below each rectangle.

1.

__________ + __________ = 1 whole

2.

__________ + __________ = 1 whole

3.

__________ + __________ = 1 whole

4.

__________ + __________ = 1 whole

Color the Fraction
用颜色表示分数

Directions: Color（涂色）the shapes（图形）to show each of the following fractions:

$\frac{3}{8}$ $\frac{7}{10}$ $\frac{3}{5}$ $\frac{2}{3}$ $\frac{3}{4}$

1.

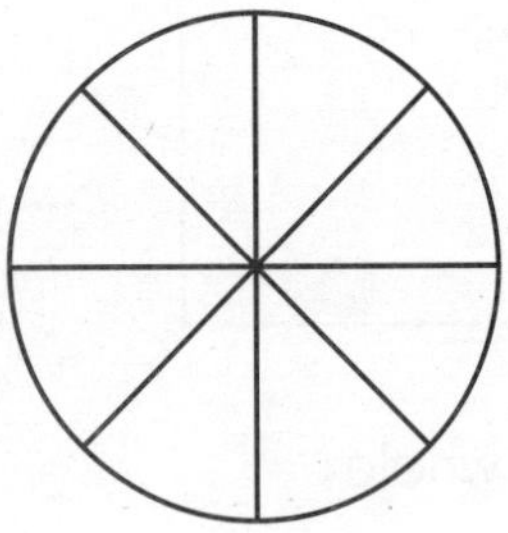

Fraction colored ______________

Fraction not colored ___________

2.

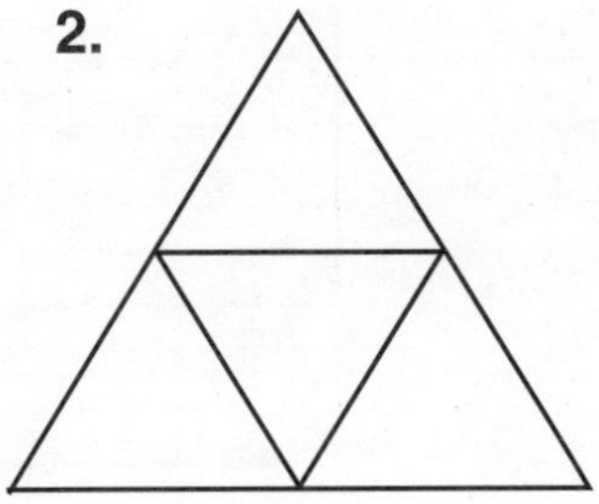

Fraction colored ______________

Fraction not colored ___________

3.

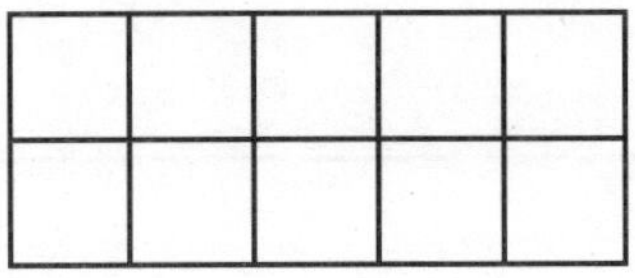

Fraction colored ______________

Fraction not colored ___________

4.

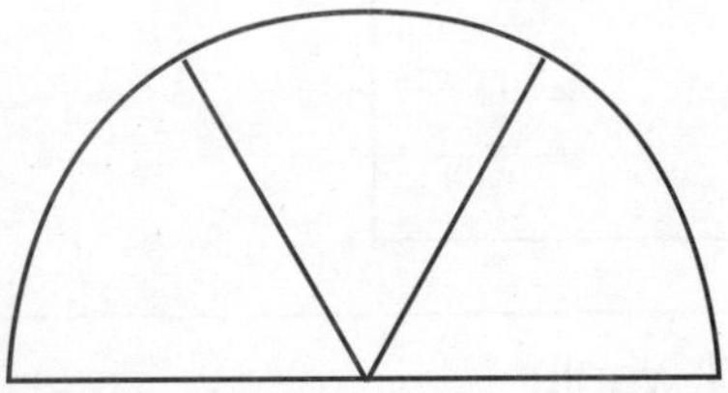

Fraction colored ______________

Fraction not colored ___________

5.

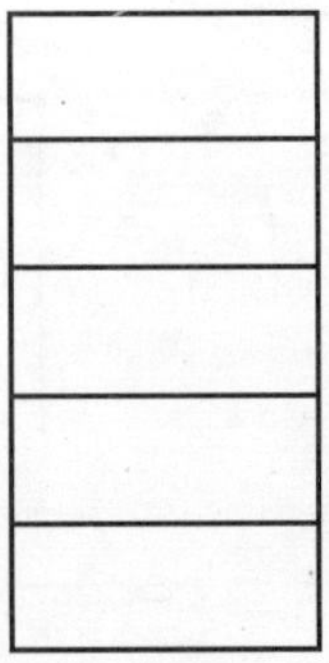

Fraction colored ______________

Fraction not colored ___________

Directions: Write the fraction shaded in each of the shapes below.

6.

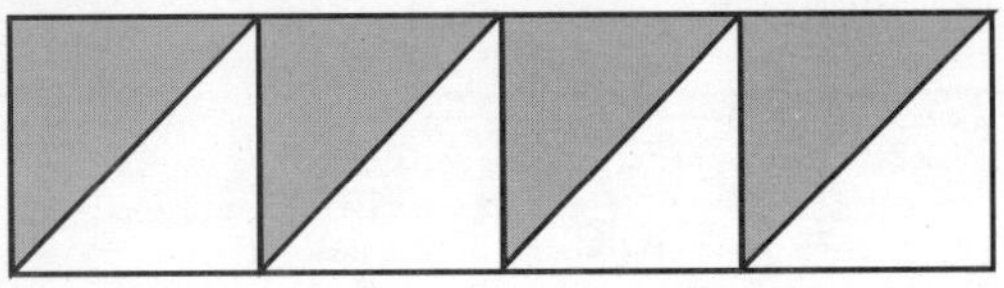

Fraction colored ______________

Fraction not colored ___________

7.

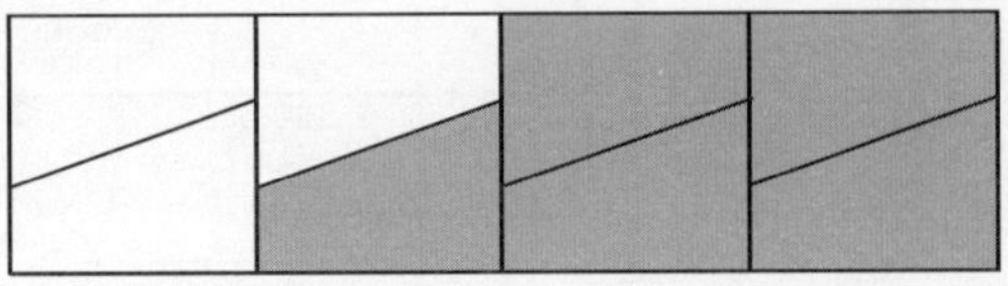

Fraction colored ______________

Fraction not colored ___________

Looking for a Pattern
找规律

Patterns can be found everywhere—in numbers, in shapes, in nature. So, it is not surprising that the strategy of "Looking for a Pattern" is used often to solve math problems. When you find a pattern, it becomes easy to predict（预测） what comes next.

You may use more than one strategy to solve a problem. As you learn more about "Looking for a Pattern," you will find that creating a table first is often the best way to see a pattern.

Read the following information to learn more about how to look for patterns when solving problems.

Problem: Get in Shape!

The Problem
The Ross family is working on a fitness program（健身计划）. On the first day, they ride bikes around the track（跑道） one time. On the second day, they ride around the track four times. On the third day, they ride seven times.

How many days must the Ross family exercise before they reach their goal（目标） of riding around the track 10 times?

Understanding the Problem
- *What do I know?*
  The number of times they ride around the track for the first, second, and third days is one, four, and seven.
- *What do I need to find out?*
  How many days must the Ross family exercise before they reach their goal?

Planning and Communicating a Solution
As you can see from the problem, the number of times around the track is increasing（增加的） each day.

You can create a table, like the one below, to see if there is a pattern.

Day	Times Around the Track
1	1
2	4
3	7

Look at the table to find the pattern.

The numbers are increasing by three times around each day. So, it will take four days for the Ross family to reach their goal of 10 times around the track.

Reflecting and Generalizing
Once the table has been created, it is easy to see a pattern.

Extension
When would they be able to ride 16 times around the track?

Standardized Test Preparation 11
标准考前测试 11

1 Round 114 to the nearest 100.

Ⓐ 200

Ⓑ 100

Ⓒ 110

Ⓓ 14

2 What fraction（分数） of the squares are black?

Ⓕ $\frac{1}{8}$

Ⓖ $\frac{2}{8}$

Ⓗ $\frac{4}{8}$

Ⓙ $\frac{8}{8}$

3 Which drink was least popular（受欢迎的）?

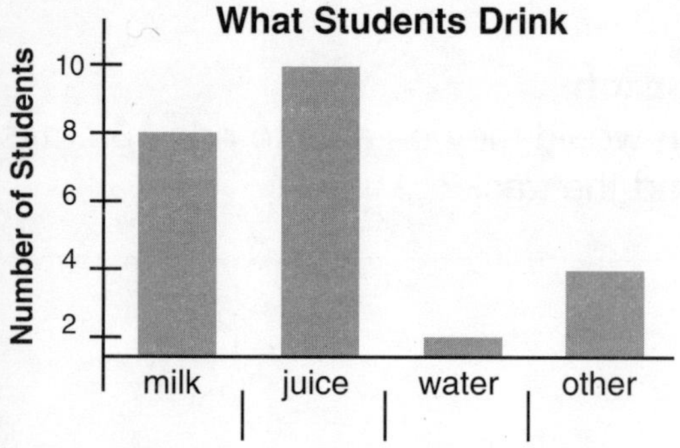

Ⓐ juice（果汁） Ⓒ water

Ⓑ milk Ⓓ other

4 About how long is a pencil?*

Ⓕ 6.0 m

Ⓖ 5.9 in.

Ⓗ 6.5 km.

Ⓙ 6.4 ft.

*本题单位换算参见附录B：数学用表。

5 Denali in Alaska（美国阿拉斯加州的德纳里峰） is twenty thousand three hundred twenty feet tall. Which number shows this height?

Ⓐ 2,302 ft.

Ⓑ 20,320 ft.

Ⓒ 2,032 ft.

Ⓓ 20,203 ft.

6 Pick one question from this test. Explain how and why you chose your answer.

____

____

____

____

____

____

Fractions of a Set

分数大集合

1. What fraction of the circles are white?

○●●●●○

A $\frac{4}{6}$ B $\frac{5}{2}$ C $\frac{2}{4}$ D $\frac{2}{6}$

2. What fraction of the circles are black?

○○●●○○○

A $\frac{5}{2}$ B $\frac{5}{7}$ C $\frac{2}{7}$ D $\frac{2}{5}$

3. What fraction of the circles are black?

○●●○○○○●○○

A $\frac{3}{10}$ B $\frac{7}{10}$ C $\frac{7}{3}$ D $\frac{3}{7}$

4. What fraction of the circles are white?

○○●

A $\frac{1}{3}$ B $\frac{1}{2}$ C $\frac{5}{3}$ D $\frac{2}{3}$

5. What fraction of the circles are white?

○○○●●○●○●○●

A $\frac{6}{11}$ B $\frac{5}{6}$ C $\frac{5}{11}$ D $\frac{6}{5}$

6. What fraction of the circles are white?

●●○○○○

A $\frac{2}{6}$ B $\frac{4}{6}$ C $\frac{2}{4}$ D $\frac{5}{3}$

7. What fraction of the circles are black?

●●●●●●●●○○○

A $\frac{8}{3}$ B $\frac{8}{11}$ C $\frac{3}{8}$ D $\frac{3}{11}$

8. What fraction of the circles are black?

○●○○○○

A $\frac{1}{6}$ B $\frac{5}{6}$ C $\frac{1}{5}$ D $\frac{5}{2}$

Picture Fractions
画出分数

1. Circle 4 equal sets（圈出4个相等的部分）.

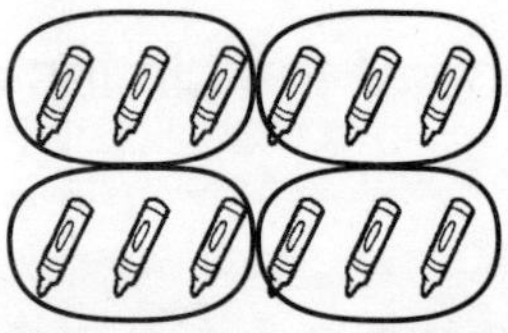

There are __3__ in each set.
Each set is __$\frac{1}{4}$__ of the whole.

2. Circle 6 equal sets.

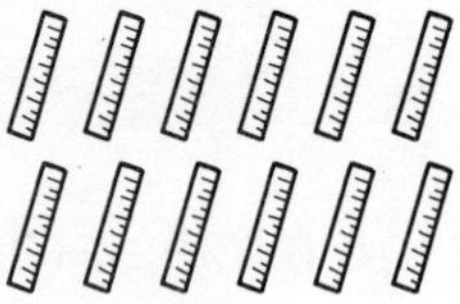

There are _______ in each set.
Each set is _______ of the whole.

3. Circle 1 set.

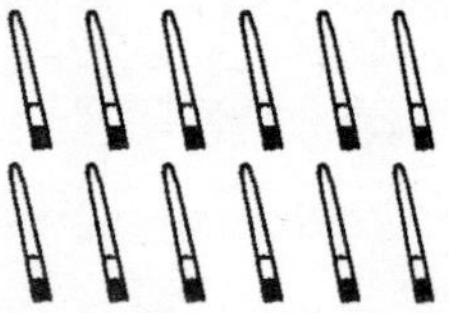

There are _______ in each set.
Each set is _______ of the whole.

4. Shade each circle to show the fraction.

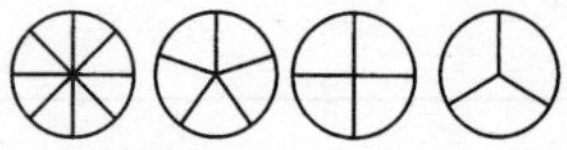

$\frac{1}{8}$ $\frac{1}{5}$ $\frac{1}{4}$ $\frac{1}{3}$

Circle the fraction that is equivalent（相当于）to $\frac{2}{8}$.

5. Shade each circle to show the fraction.

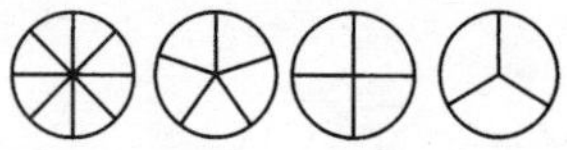

$\frac{1}{8}$ $\frac{1}{5}$ $\frac{1}{4}$ $\frac{1}{3}$

Circle the fraction that is larger than $\frac{1}{4}$.

6. Circle 3 equal sets.

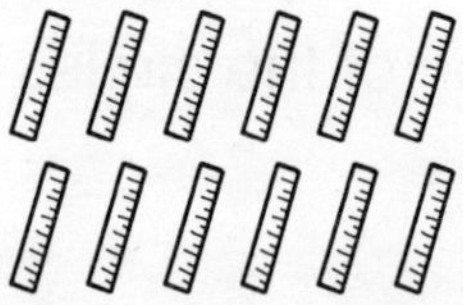

There are _______ in each set.
Each set is _______ of the whole.

7. Circle 2 equal sets.

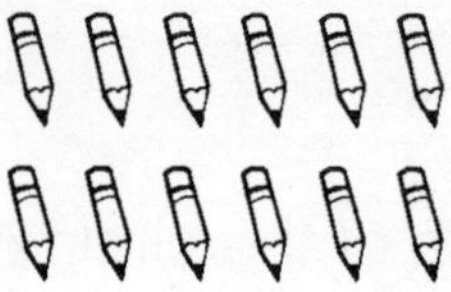

There are _______ in each set.
Each set is _______ of the whole.

8. Shade each circle to show the fraction.

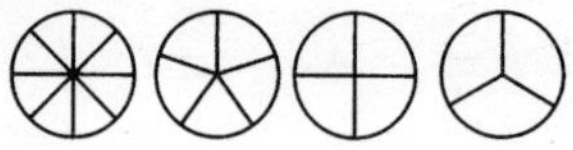

$\frac{1}{8}$ $\frac{1}{5}$ $\frac{1}{4}$ $\frac{1}{3}$

Draw a line under the smallest fraction. Make an *X* on the largest fraction.

9. Circle 2 equal sets.

There are _______ in each set.
Each set is _______ of the whole.

10. Circle 4 equal sets.

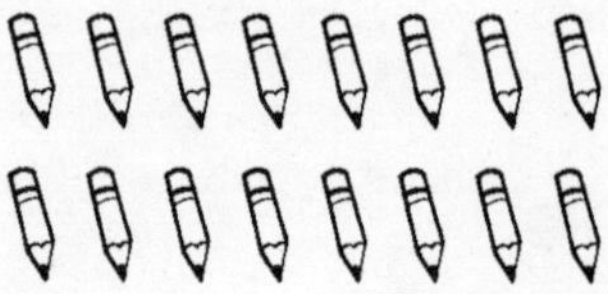

There are _______ in each set.
Each set is _______ of the whole.

Taking a Field Trip
实地考察旅行中的数学

"I have a surprise for you boys and girls," Mrs. Mally said with a great big grin（咧嘴笑）on her face. "Next week we will take a field trip to a museum（博物馆）." The children in Room 8 cheered（欢呼）.

"I love going on field trips!" exclaimed（大叫）Louie.

Mrs. Mally explained（解释）to the children that it is a lot of work to get ready for a field trip. "I would like you to help me make some of the arrangements（准备）," she said.

"First, we have to make reservations（预约）with the museum. We have to let the museum know what day we will be coming and how many people we will bring with us. The other four third grade classrooms will be joining us. Also, we will need to count adults（成年人）. Who would like to help me select a day and count how many people we will bring with us?" Sophia and Luca raised（举起）their hands.

"How are we going to get there?" questioned Mark.

"I was just getting to that," said Mrs. Mally. "We are going to get buses to take us to the museum. Who would like to help figure out（算出）how many buses we will need?" Mrs. Mally selected Ty and Kim to help with this task（任务）.

"We are going to need permission（许可）from your parents," said Mrs. Mally. "I will copy permission slips（许可协议）for each of you. Which children can work together to write a letter to the parents to let them know what we are going to do?" Sevan and Melia volunteered（自愿地，自告奋勇地）to write the letter.

"Finally, I think it would be nice to find out a little bit of information about the kinds of things we can see when we get to the museum," said Mrs. Mally.

"Hey, Nikki and I can jump on the Internet（上网）and look up the museum's website（网站）. We'll see if we can find any information!" exclaimed Cho.

Mrs. Mally looked pleased. "Well, it looks like we will be ready for our field trip next week, thanks to your help!" she said.

Taking a Field Trip *(cont.)*

What Is the Problem?

The third-grade students at Mayfield Elementary School are going on a field trip to a museum. There are five third-grade classrooms with 20 students in each class. In addition to the teacher（加上老师）, each class needs to take four adults on the trip. Your job is to help make the arrangements for the field trip.

Use the information above and in the problems below to answer the questions. Before you begin solving each problem be sure to locate the information you will need. Think about how you will organize the information in order to look for a pattern.

Remember: If you have trouble planning ways to solve a problem, reread it and rethink your strategy. When you have a solution, think about whether your answer makes sense.

Problem-Solving Strategy: Looking for a Pattern

Problem A

How many adults will need to go on the field trip? After you create your table, what is the multiplication pattern（乘法规律）that emerges（显现）?

Hint: Do not forget the teacher for each class!

Problem B

How many students will go on the field trip? How many total people will be going on the field trip?

Hint: Reread the problem to find out how many students are in each classroom.

Problem C

A bus can hold（容纳）60 people. How many buses will be needed for this field trip?

Hint: Fill a bus completely before using another bus（一辆坐满，再用下一辆）.

Extension

Look at your table and your answer for Problem C. Is there room left on the final bus? If a fourth grade class of 20 students (with the necessary adults) also wanted to go on the field trip, would there be room?

Each class at the museum will be given a bag of 23 souvenirs（纪念品）to take home after they have visited the museum. How many souvenirs will be given to all five classrooms total? How many extras（额外的，多出来的）will they have after each student has been given a souvenir?

Standardized Test Preparation 12

标准考前测试 12

1 What is the next number in the sequence?

3, 6, 9, 12, ____

Ⓐ 13

Ⓑ 15

Ⓒ 17

Ⓓ 19

2 Which operations are inverse (相反的)?

Ⓕ addition (加法) and subtraction (减法)

Ⓖ addition and multiplication (乘法)

Ⓗ subtraction and division (除法)

Ⓙ division and addition

3 Sasha has 6 flowers. She gives an equal number of flowers to 2 friends. How many flowers does each friend get?

Ⓐ 8

Ⓑ 12

Ⓒ 4

Ⓓ 3

4 What fraction of the shapes are white?

Ⓕ $\frac{1}{4}$

Ⓖ $\frac{2}{4}$

Ⓗ $\frac{3}{4}$

Ⓙ $\frac{4}{4}$

5 Which is the best estimate of the time it would take to wash a load of clothes in the washing machine (用洗衣机洗一缸衣服的时间)?

Ⓐ 30 seconds

Ⓑ 30 minutes

Ⓒ 30 hours

Ⓓ 30 days

6 Pick one question from this test. Explain how and why you chose your answer.

____

____

____

____

____

____

Fraction Wall
分数墙

Directions: Write the fraction for each "brick（砖）" below. The first one has been done for you.

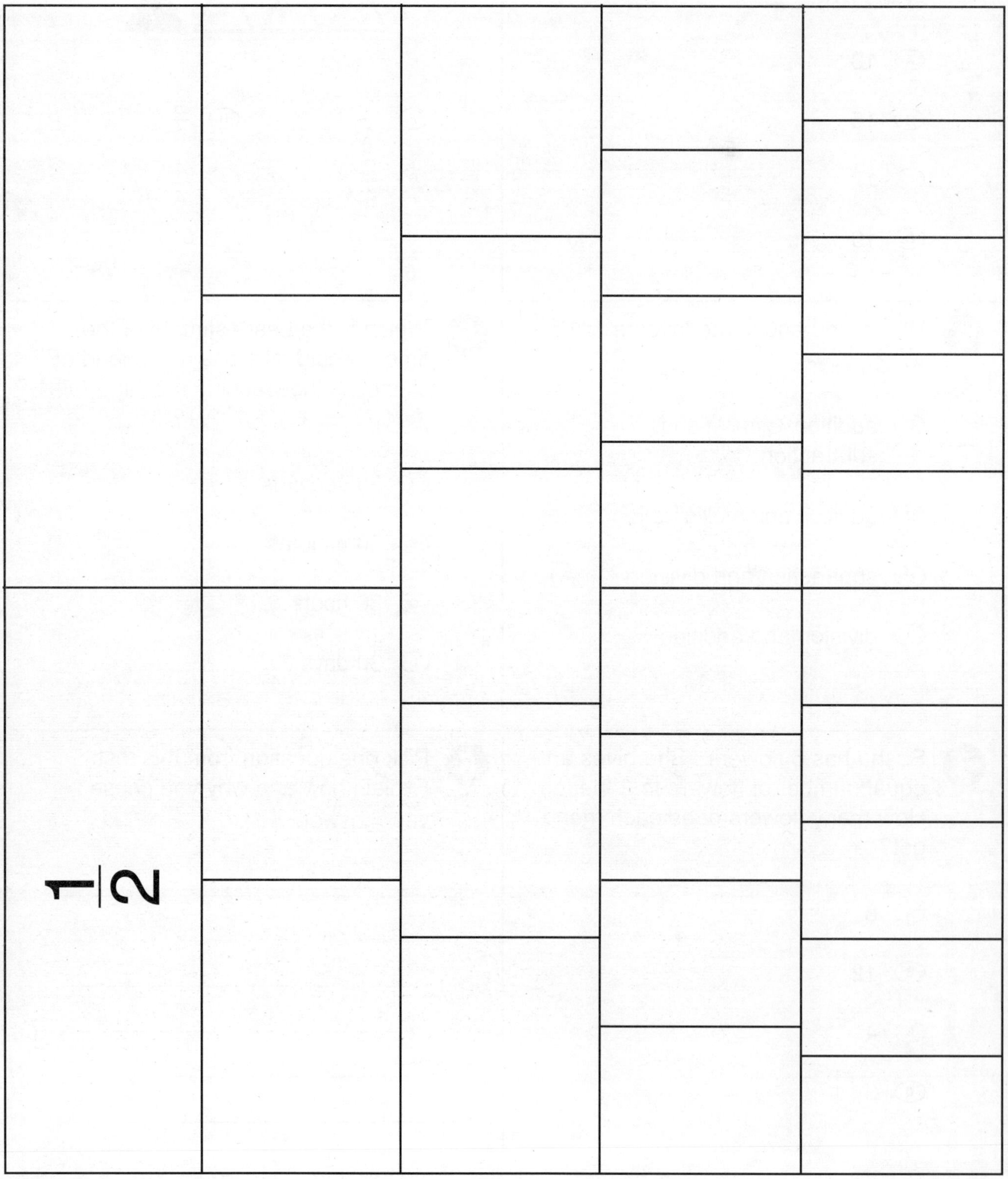

Greater Than or Less Than
大于还是小于

Directions: Shade the boxes to match（匹配） the fractions. Then, compare（比较） the fractions and write > or <. The first one has been done for you.

1.

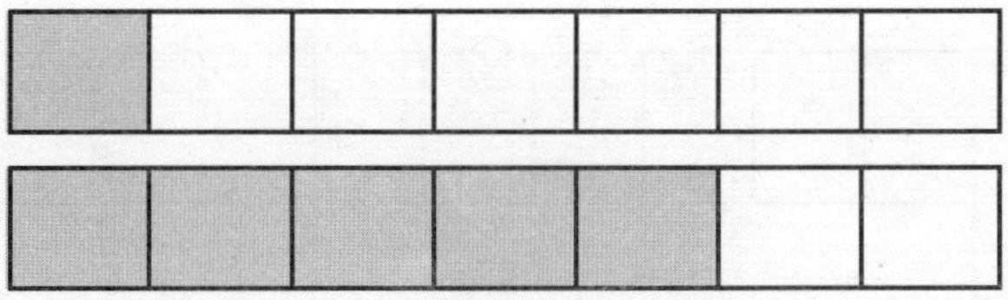

$\frac{1}{7}$ $<$ $\frac{5}{7}$

2.

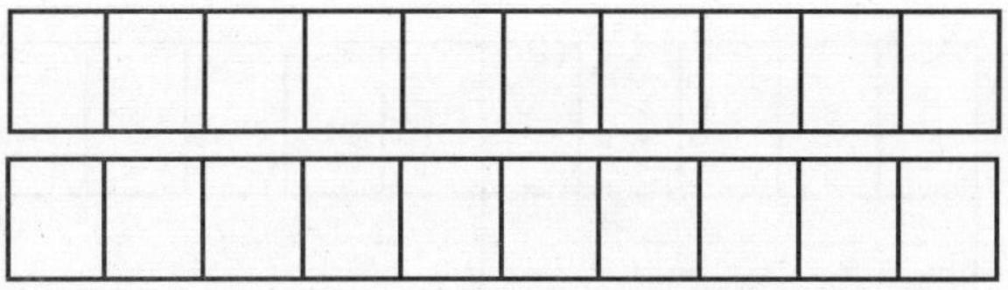

$\frac{2}{10}$ ○ $\frac{5}{10}$

3.

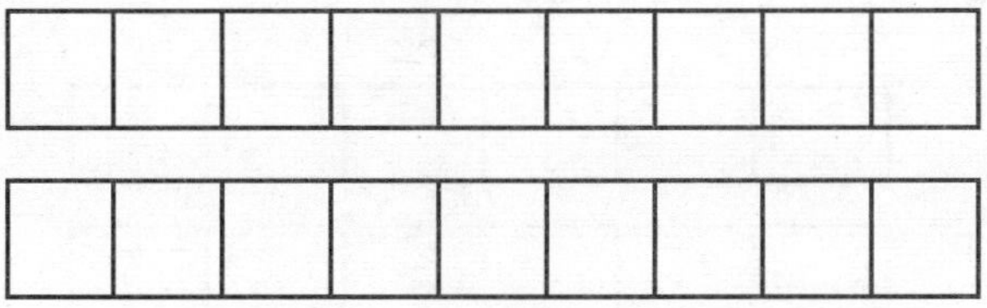

$\frac{7}{9}$ ○ $\frac{4}{9}$

4.

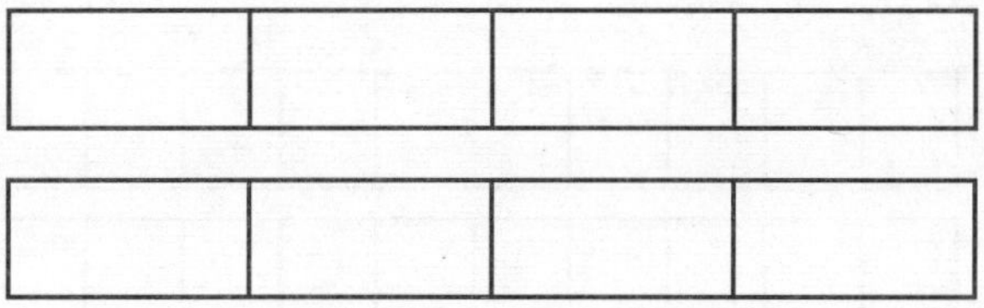

$\frac{3}{4}$ ○ $\frac{1}{4}$

5.

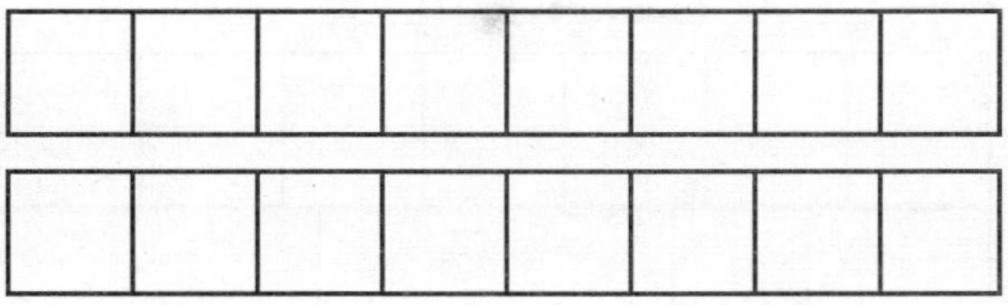

$\frac{8}{8}$ ○ $\frac{7}{8}$

6.

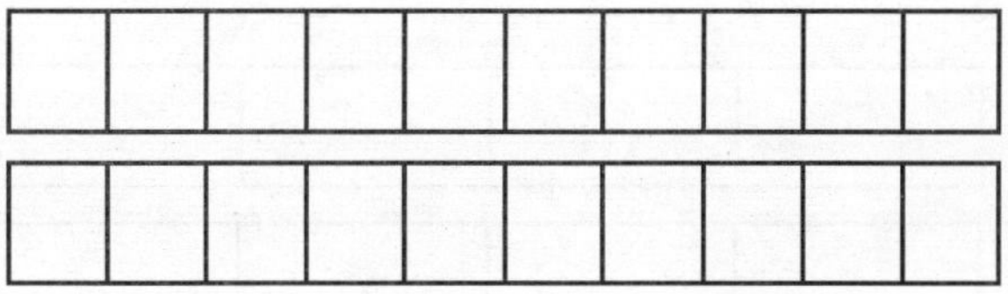

$\frac{5}{10}$ ○ $\frac{8}{10}$

7.

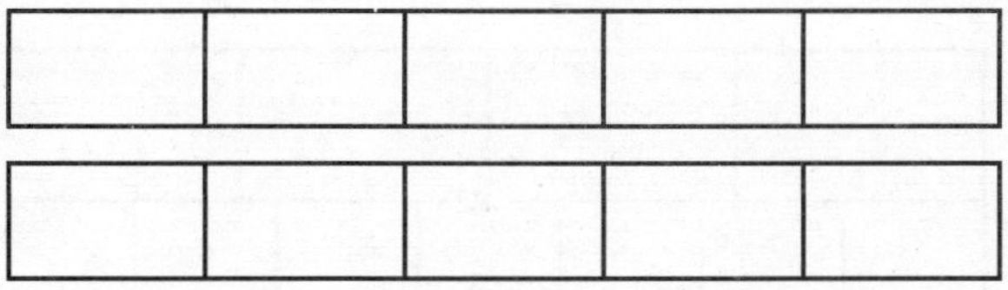

$\frac{2}{5}$ ○ $\frac{3}{5}$

8.

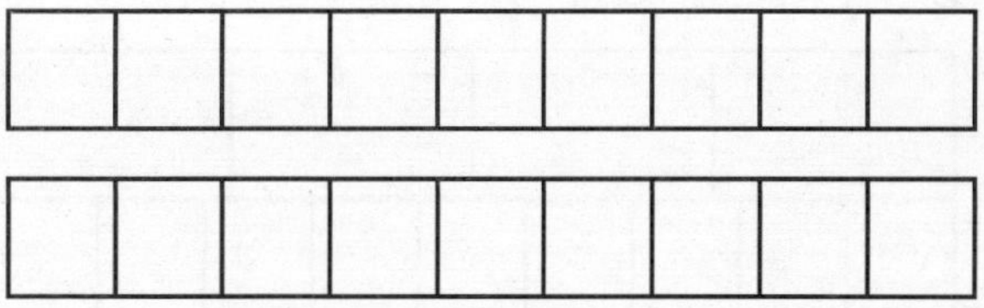

$\frac{6}{9}$ ○ $\frac{7}{9}$

Comparing Fractions

分数比大小

Directions: Shade the boxes to match the fractions. Then, compare the fractions and write > or <.

1.

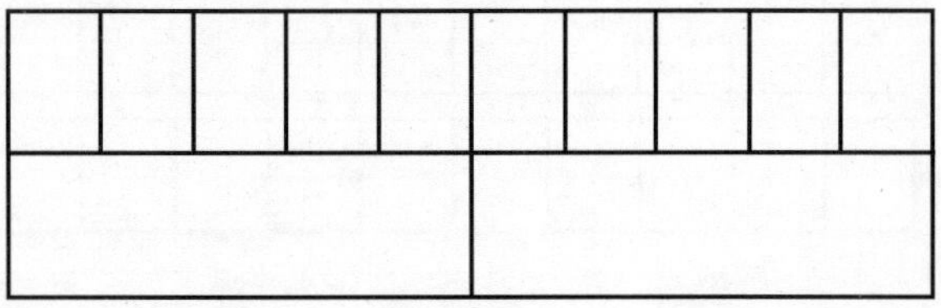

$\frac{1}{10}$ ◯ $\frac{1}{2}$

2.

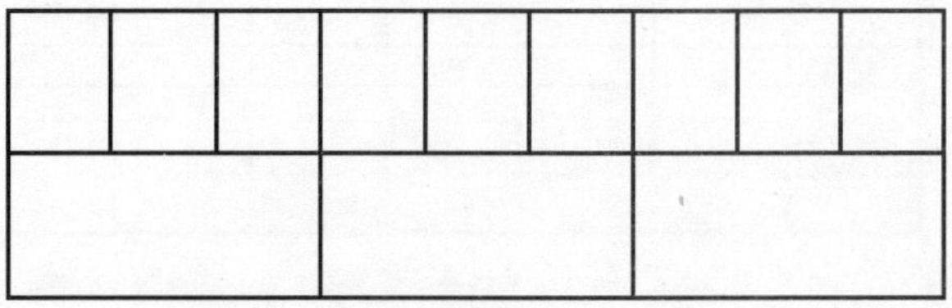

$\frac{2}{9}$ ◯ $\frac{2}{3}$

3.

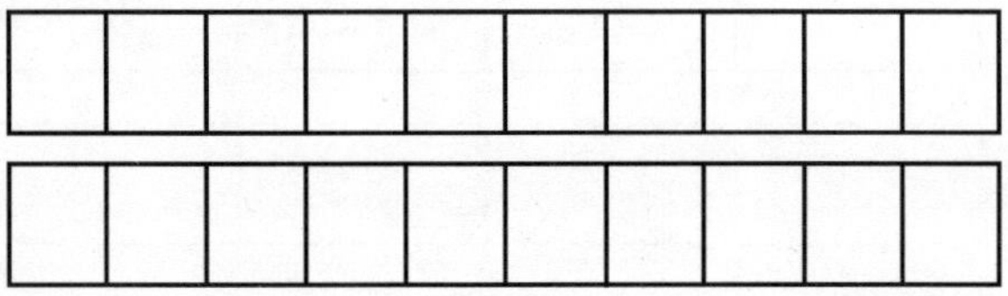

$\frac{6}{10}$ ◯ $\frac{4}{10}$

4.

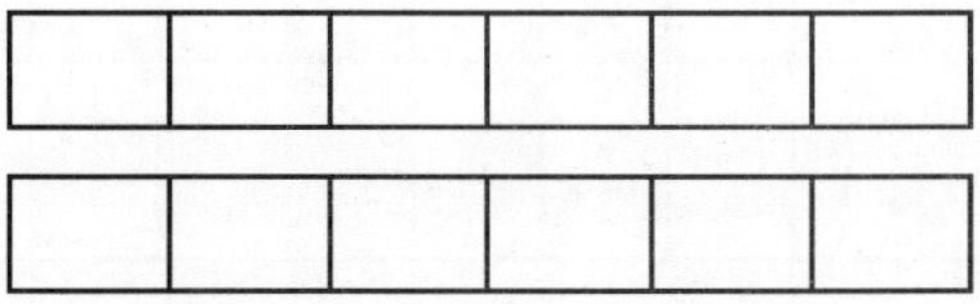

$\frac{3}{6}$ ◯ $\frac{1}{6}$

5.

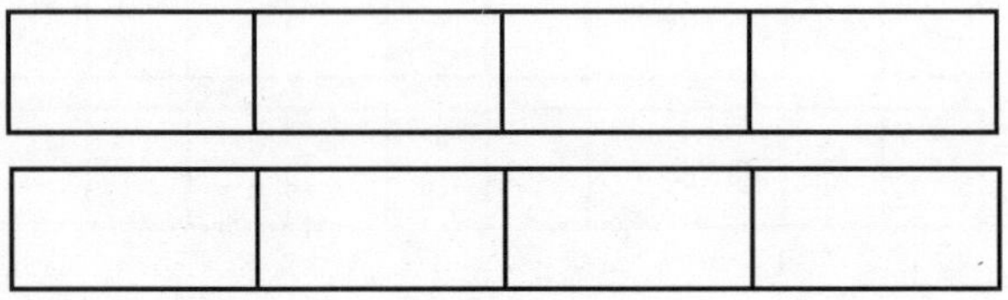

$\frac{3}{4}$ ◯ $\frac{2}{4}$

6.

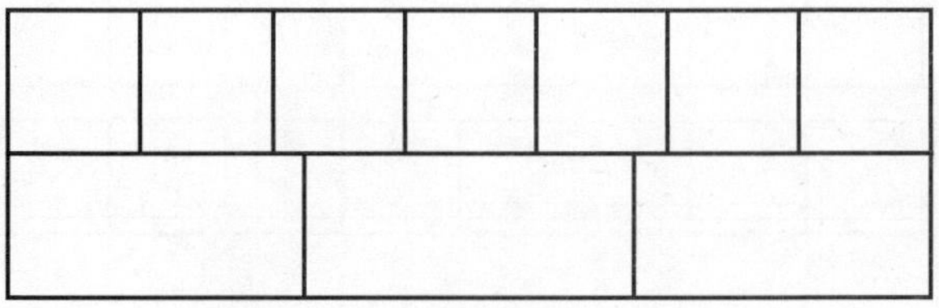

$\frac{1}{7}$ ◯ $\frac{1}{3}$

7.

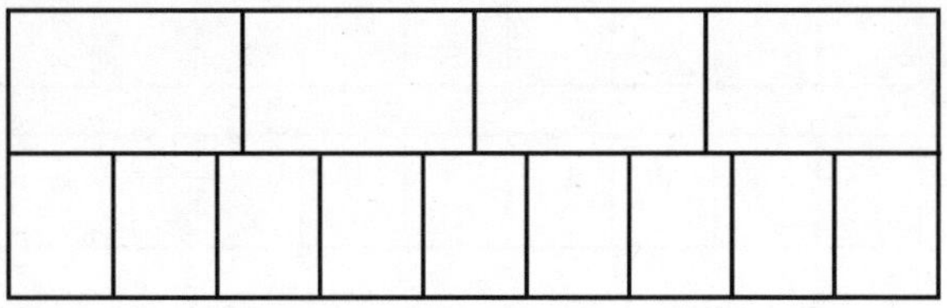

$\frac{1}{4}$ ◯ $\frac{1}{9}$

8.

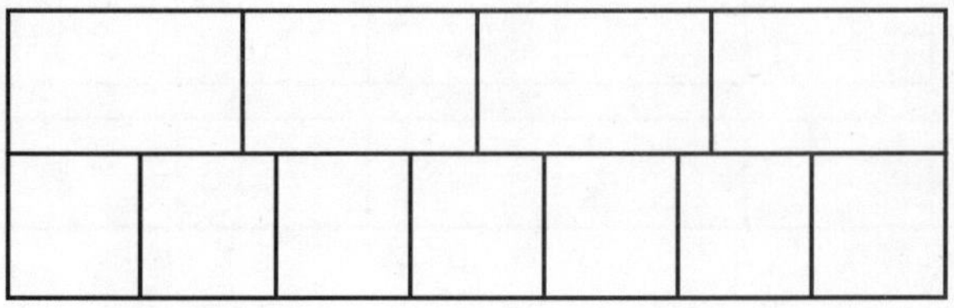

$\frac{3}{4}$ ◯ $\frac{3}{7}$

Looking for a Pattern Problems

找规律 练习

Directions: Solve each problem and show your work.

Problem A

Extend（延伸）each pattern. Write the rule（规则）.

65, 75, 85, ______, ______, ______, ______

Rule: ____

189, 289, 389, ______, ______, ______, ______

Rule: ____

413, 418, 423, ______, ______, ______, ______

Rule: ____

Problem B

Extend the pattern. Write the rule.

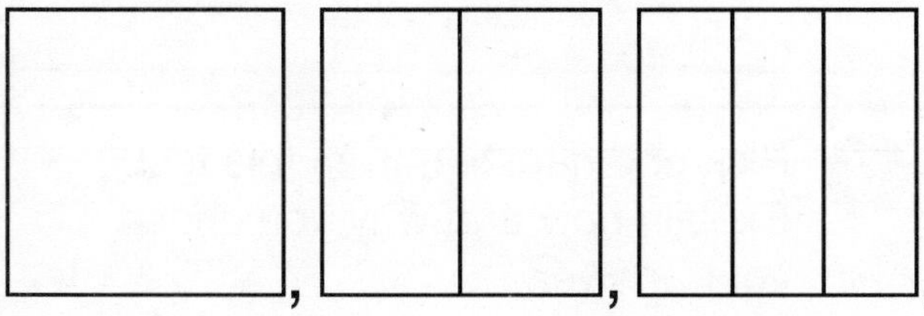, ______________, ______________

Rule: ____

Standardized Test Preparation 13
标准考前测试 13

1 Round 139 to the nearest 100.

Ⓐ 200

Ⓑ 100

Ⓒ 140

Ⓓ 14

2 Which line is about 4.5 cm long?

Ⓕ 

Ⓖ 

Ⓗ 

Ⓙ 

3 Which piece of equipment won the most votes?（哪项设备得票最多）

Favorite Playground Equipment

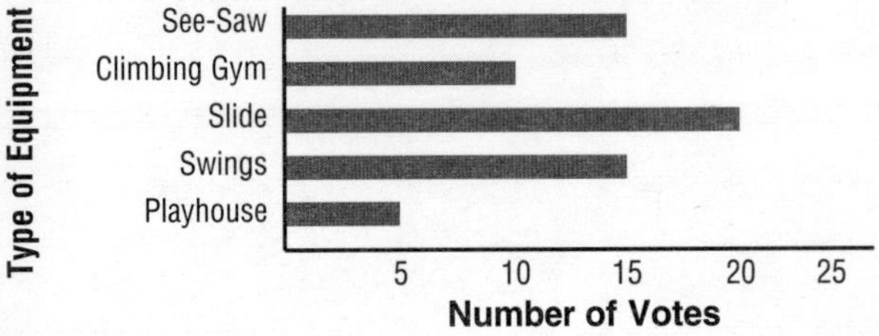

Ⓐ slide（滑梯）

Ⓑ climbing gym（攀岩）

Ⓒ swings（秋千）

Ⓓ see-saw（跷跷板）

4 Which fraction is greater than $\frac{1}{2}$?

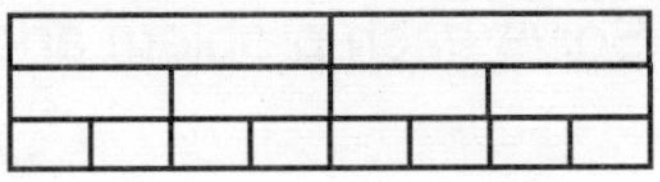

Ⓕ $\frac{5}{8}$

Ⓖ $\frac{1}{4}$

Ⓗ $\frac{2}{4}$

Ⓙ $\frac{3}{8}$

5 Which player had the lowest score（得分）?

Points Scored in Math Contest

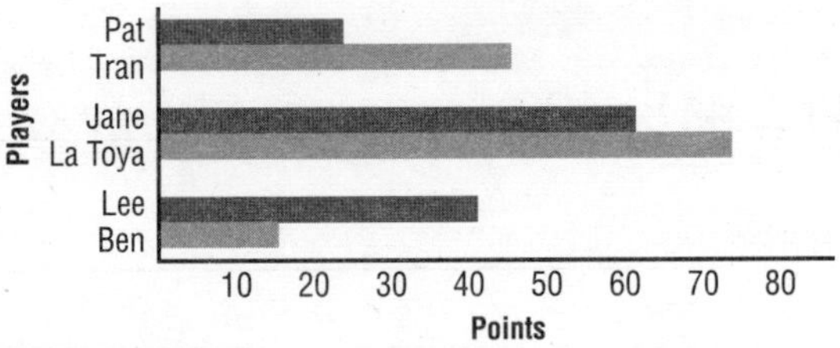

Ⓐ Ben

Ⓑ Tran

Ⓒ La Toya

Ⓓ Lee

6 Pick one question from this test. Explain how and why you chose your answer.

____

____

____

____

____

____

Lesson 14

Equivalent Fractions

相等的分数

Directions: Use the fraction walls to make equivalent fractions.

1\.

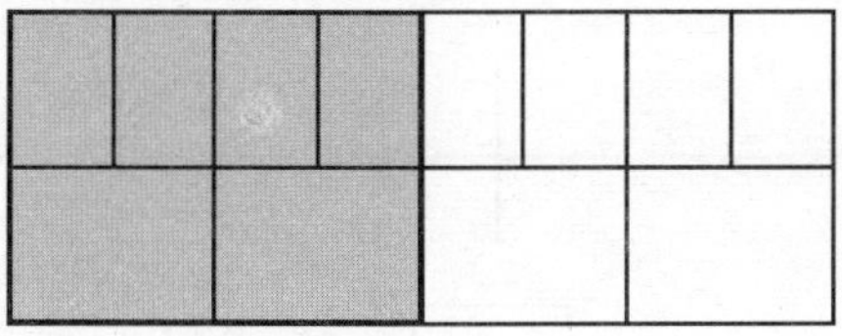

$\frac{4}{8} = \frac{\square}{4}$

2\.

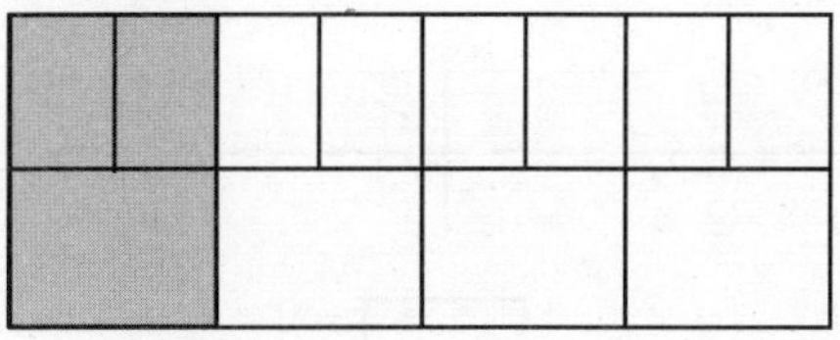

$\frac{2}{8} = \frac{\square}{4}$

3\.

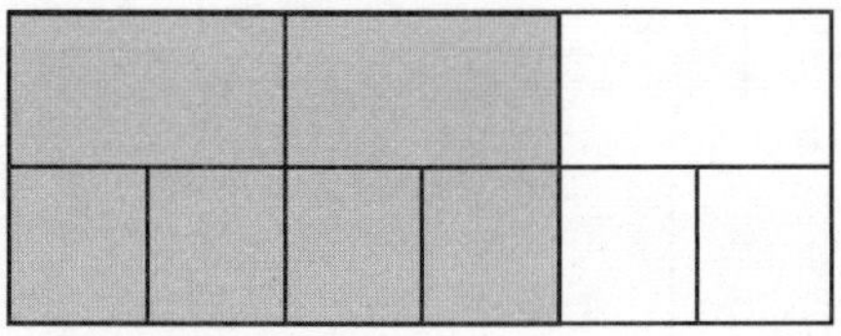

$\frac{\square}{3} = \frac{4}{6}$

4\.

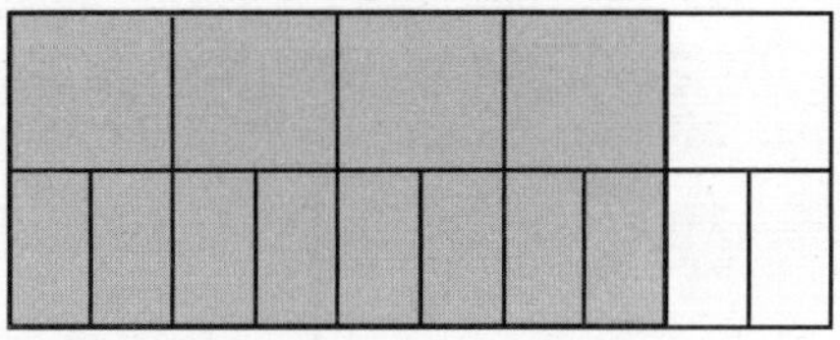

$\frac{\square}{5} = \frac{8}{10}$

5\.

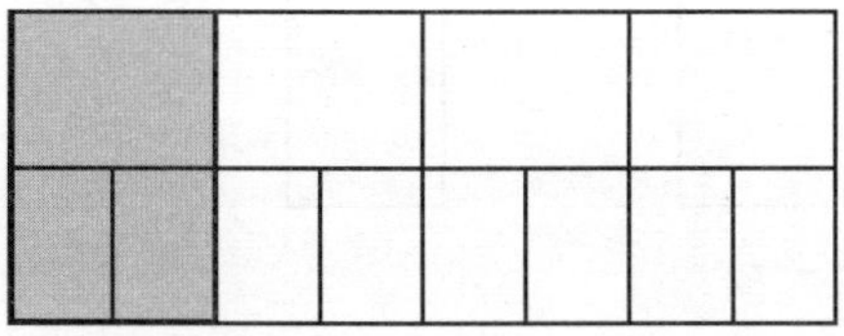

$\frac{1}{4} = \frac{\square}{8}$

6\.

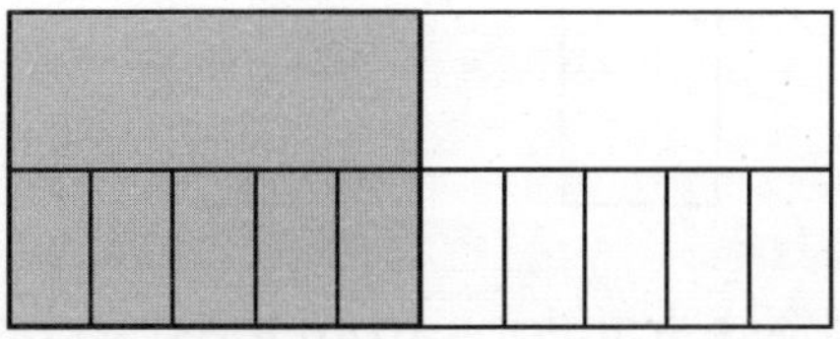

$\frac{\square}{2} = \frac{5}{10}$

7\.

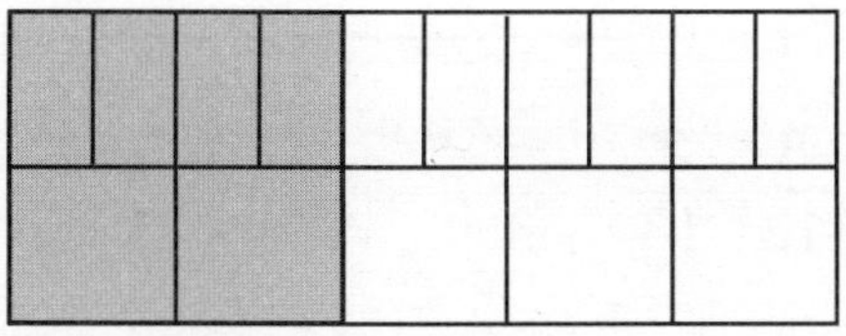

$\frac{\square}{10} = \frac{2}{5}$

8\.

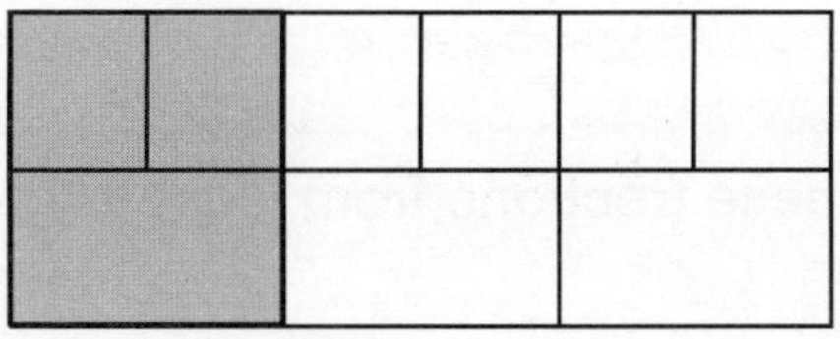

$\frac{2}{6} = \frac{\square}{3}$

Ordering Fractions

分数排序

Directions: Write the fraction that belongs in each box.

1.

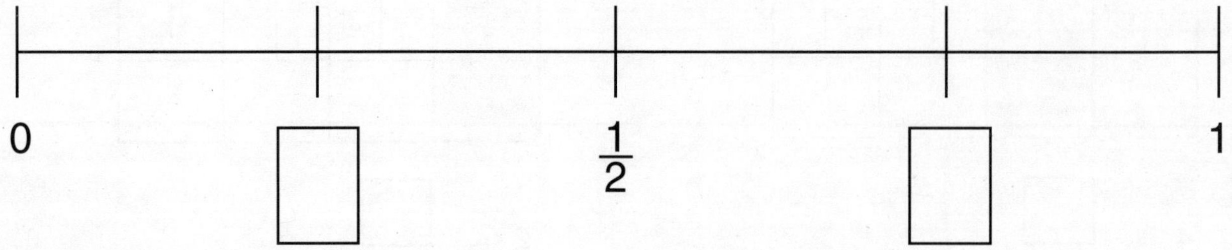

2.

3.

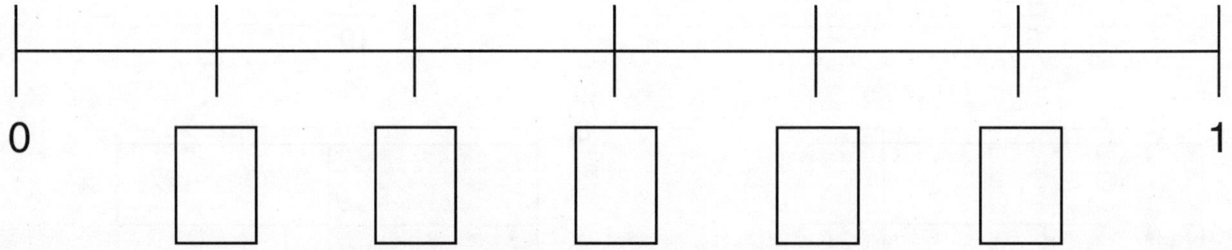

4. **Order these fractions from least to greatest:** $\frac{2}{7}, \frac{7}{7}, \frac{6}{7}, \frac{5}{7}$

____

5. **Order these fractions from least to greatest:** $\frac{11}{11}, \frac{3}{11}, \frac{5}{11}, \frac{1}{11}$

____

6. **Order these fractions from least to greatest:** $\frac{3}{13}, \frac{13}{13}, \frac{8}{13}, \frac{2}{13}$

____

7. **Order these fractions from least to greatest:** $\frac{3}{5}, \frac{5}{5}, \frac{2}{5}, \frac{1}{5}$

____

Pay Day
发薪日

"Dad, can you buy me a new MP3 player?" whined（哼哼唧唧地说）Joey.

Dad rolled his eyes（转动眼睛）. "Every time I turn around（手头富裕）, you are asking me to buy you something, Joey. Do you think I am made out of money?"

"No, but I really want a new MP3 player!" Joey continued to whine.

Dad took a deep breath（深呼吸）. "I think it is about time you started learning about money. I am going to begin giving you an allowance（零用钱） for the chores（家庭杂务） you do around the house."

Joey smiled as he said, "All right! This is going to be fantastic! Now I can buy an MP3 player."

Dad thought for a minute and then he said, "How does $15 a month sound?"

"It sounds great, but the MP3 player I want costs way more than that," complained（抱怨）Joey.

"Well, that's where a money plan comes in. You see, if you want to buy an item（一件商品或物品） that costs a lot of money, you will need a plan for how to get the money for it," Dad explained.

"What sort of plan?" asked Joey.
Dad continued, "I recommend（推荐） using some of your allowance each month for spending money and saving some of the money for something more expensive（贵的）."

"Like an MP3 player?" interrupted（打断，插嘴）Joey.

"Like an MP3 player," said Dad. "If you plan and save, you will have the MP3 player in no time（马上） and you will have some spending money, too."

"That sounds（听起来） great! I think I am going to work on my plan right now. Will you help me, Dad?" asked Joey.

"Of course, son," replied Dad.

Pay Day *(cont.)*

What Is the Problem?

Joey's dad is going to begin giving him $15 in allowance each month. Joey and his dad decide on a saving plan (存款计划): For each $2 Joey spends, he wants to save $3. The MP3 player costs $45.

Use the information above and in the problems below to answer the questions. Before you begin solving each problem, be sure to locate the information you will need. Think about how you will organize the information in order to look for a pattern.

Remember: If you have trouble planning ways to solve a problem, reread it and rethink your strategy. When you have a solution, think about whether your answer makes sense.

Problem-Solving Strategy: Looking for a Pattern

Problem A

How much money will Joey be able to save each month?

Hint: The total amount he spends and saves must equal his total allowance for the month.

Problem B

How much money will Joey be able to save in three months?

Problem C

For how many months will Joey have to use his saving plan to be able to spend his savings on the MP3 player?

Hint: This money is only from his savings from each month. He spent the spending money from each month on other things.

Extension

If Joey continues to use his savings plan, how much spending money will he have earned (赚得) in 6 months? How much will he have in his savings after 6 months if he has bought the MP3 player?

Standardized Test Preparation 14

标准考前测试 14

1 Look at the shaded part of each figure. Which figure shows $\frac{3}{4}$ shaded?

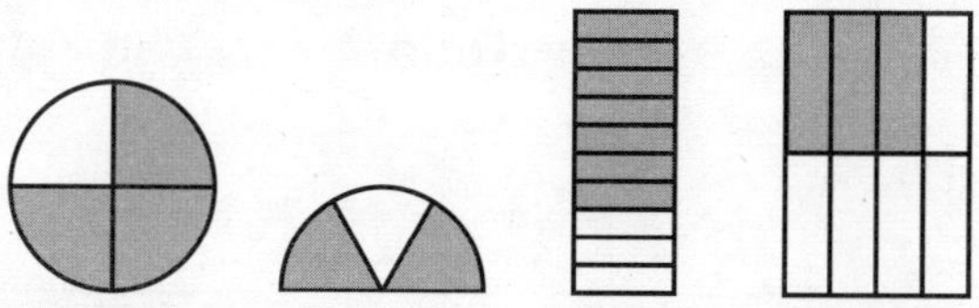

Figure 1 Figure 2 Figure 3 Figure 4

Ⓐ Figure 1

Ⓑ Figure 2

Ⓒ Figure 3

Ⓓ Figure 4

2 At the movie theater（电影院）, a soft pretzel（椒盐脆饼） costs $5.50. A bag of popcorn（爆米花） costs half as much as a soft pretzel. Ms. Chavez and her grandchildren buy 2 soft pretzels and 3 bags of popcorn. How much do they spend together?

Ⓕ $22.00

Ⓖ $19.25

Ⓗ $12.25

Ⓙ $7.75

3 8 x 6 = _____

Ⓐ 86

Ⓑ 48

Ⓒ 40

Ⓓ 2

4 What is the next number in the sequence?

30, 25, 20, 15, ____

Ⓕ 5

Ⓖ 35

Ⓗ 10

Ⓙ 0

5 Which would best be measured in minutes（以分钟计算）?

Ⓐ flying across the country（飞越一个国家）

Ⓑ eating lunch

Ⓒ driving across California（驶过加利福尼亚州）

Ⓓ painting a bedroom（粉刷卧室）

6 Pick one question from this test. Explain how and why you chose your answer.

____

____

____

____

____

____

Lesson 15

Patterns and Sequences

规律和序列

Directions: Complete each pattern and explain the rule.

Rule

1. 0, 2, 4, 6, ________ , ________ ______________________________

2. 50, 45, 40, 35, ________ , ________ ______________________________

3. 22, 32, 42, 52, ________ , ________ ______________________________

4. 601, 501, 401, ________ , ________ ______________________________

5. 41, 36, 31, 26, ________ , ________ ______________________________

6. 98, 96, 94, 92, ________ , ________ ______________________________

7. 12, 17, 22, 27, ________ , ________ ______________________________

Directions: Study the sequence of shapes in each pattern. Then, draw the next shape that would follow in each series（系列）.

8. 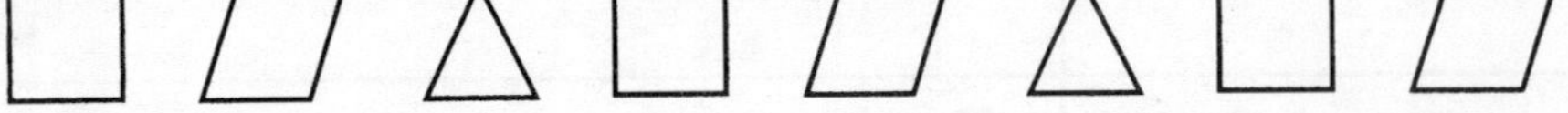 ______________

9.  ______________

10.  ______________

Looking for a Pattern Group Problems

找规律 分组练习

Directions: Solve Problem A and show your work. Using the strategy you've studied this week, write and solve your own problems for Problem B and Problem C.

Problem A

Extend the pattern for each set of fractions.

$\frac{1}{2}$, $\frac{1}{3}$, $\frac{1}{4}$, $\frac{1}{5}$, ________, ________, ________, ________

$\frac{1}{1}$, $\frac{1}{2}$, $\frac{1}{4}$, $\frac{1}{8}$, ________, ________, ________, ________

$\frac{1}{2}$, $\frac{2}{4}$, $\frac{3}{6}$, $\frac{4}{8}$, ________, ________, ________, ________

Problem B: ________________________

____

____

____

____

____

Show work:

Answer: ____________________________

Problem C: ________________________

____

____

____

____

____

Show work:

Answer: ____________________________

Standardized Test Preparation 15
标准考前测试 15

1 173 + ______ = 180

Ⓐ 7

Ⓑ 10

Ⓒ 17

Ⓓ 25

2 There are 5 people in Carlota's group. Each person needs 4 sheets of paper (4页纸). How can Carlota figure out how many sheets of paper her group will need in all?

Ⓕ 5 + 4 = 9 sheets

Ⓖ 5 – 4 = 1 sheets

Ⓗ 5 x 4 = 20 sheets

Ⓙ 5 + 4 + 1 = 10 sheets

3 Which fraction is equal to $\frac{1}{2}$?

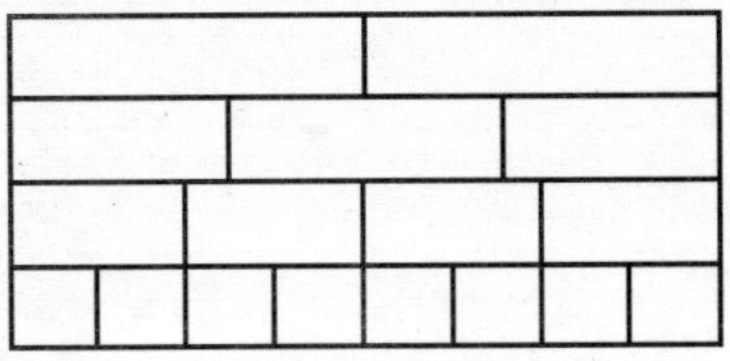

Ⓐ $\frac{1}{3}$

Ⓑ $\frac{2}{3}$

Ⓒ $\frac{2}{4}$

Ⓓ $\frac{3}{8}$

4 Lucas bought 45 tomato plants for his garden. He wants to put them in 5 rows (行, 排). How many tomato plants can he plant in each row?

Ⓕ 1 plants

Ⓖ 5 plants

Ⓗ 9 plants

Ⓙ 10 plants

5 If a square has sides that are 4 cm long, what is its perimeter (周长)?

Ⓐ 4 cm

Ⓑ 8 cm

Ⓒ 12 cm

Ⓓ 16 cm

6 Pick one question from this test. Explain how and why you chose your answer.

____

____

____

____

____

____

Lesson 16

Robot
测量机器人

Directions: Measure the robot below in centimeters, recording all the measurements you can. Measure to the nearest 0.5 centimeter.

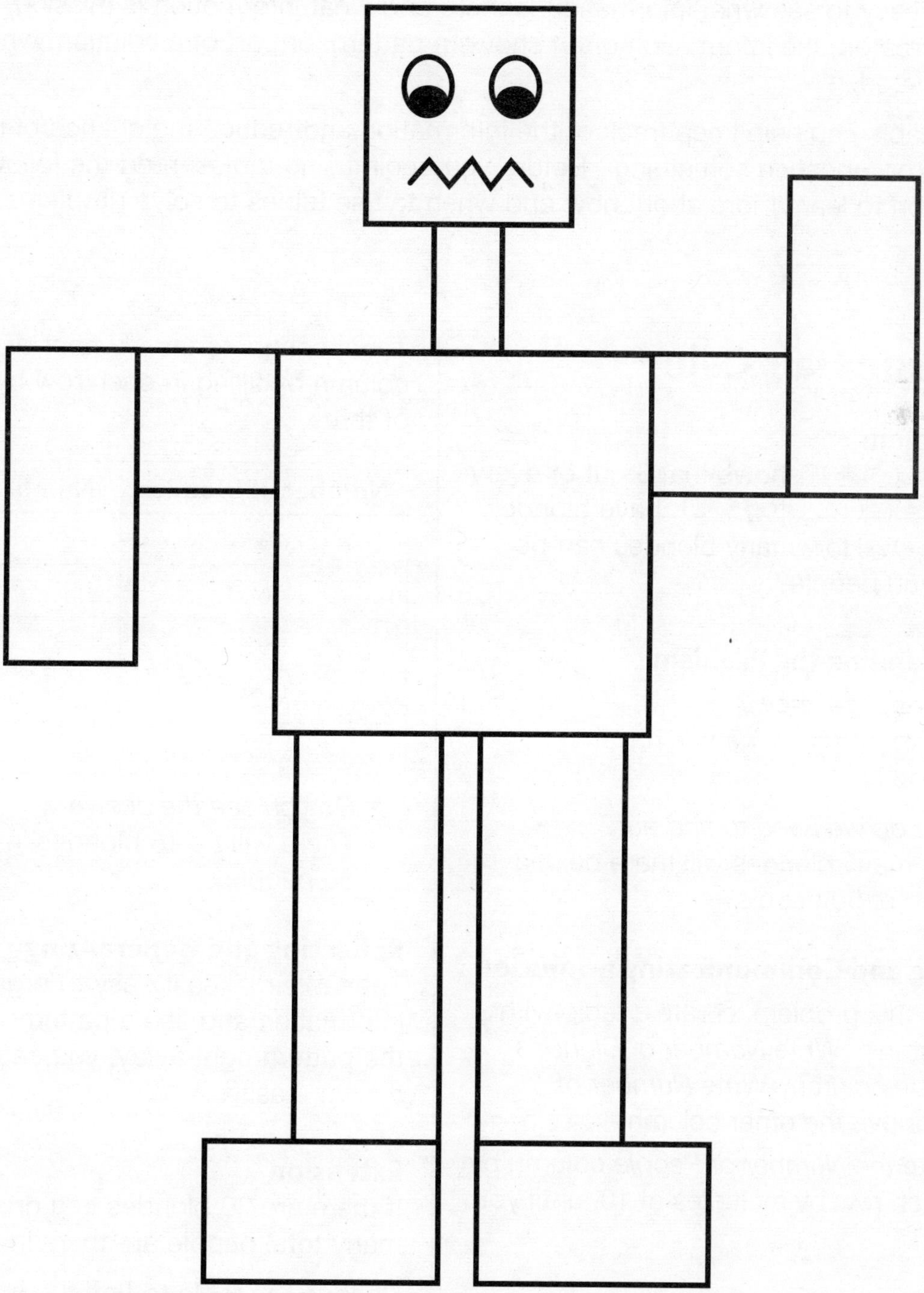

Challenge

Directions: On another piece of paper, draw a robot dog using only straight lines（仅用直线）. Write the measurements for it.

Creating a Table 2

制作表格 2

When a problem has lots of information, placing the information in a table is a good idea. A table helps you organize the information so that it can be easily understood. A table makes it easy to see what information is there and what information is missing. When a table is created, the information often shows a pattern, or part of a solution, which can then be completed.

Using a table can help keep track of the information and reduce the chance of making mistakes or repeating something. Before you begin using tables, read the following information to learn more about how and when to use tables to solve problems.

Problem: Hair Color

The Problem

Research（调查） shows that 3 out of every 10 people（每10人中有3人） have blonde hair（金发）. How many blondes can be found in 50 people?

Understanding the Problem

- *What do we know?*
  3 out of every 10 people have blonde hair.
- *What do we need to find out?*
  How many blondes will there be in a group of 50 people?

Planning and Communicating a Solution

To solve this problem, create a table with two columns. Write *Number of Blondes* above one column. Write *Number of People* above the other column.

Complete the *Number of People* column by filling each row by multiples of 10, until you get to 50.

Then, complete the *Number of Blondes* column by filling in each row by multiples of three.

Number of Blondes	Number of People
3	10
6	20
9	30
12	40
15	50

- *Do you see the answer?*
  There will be 15 blondes in a group of 50 people.

Reflecting and Generalizing

The table makes it easy to organize the information and see a pattern. If you see the pattern right away, you can find the answer easily.

Extension

If there are 30 blondes in a group, how many total people are there in the group?

Change the table to find the answer.

Standardized Test Preparation 16

标准考前测试 16

1

$$\begin{array}{r} 54 \\ +\ 38 \\ \hline \end{array}$$

Ⓐ 81

Ⓑ 82

Ⓒ 92

Ⓓ 101

2 Which number sentence is true?

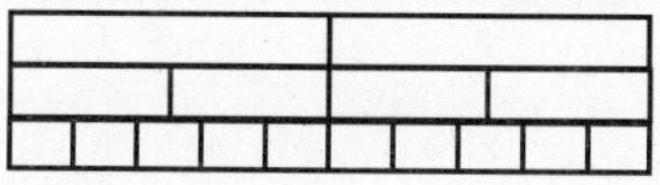

Ⓕ $\frac{1}{2} > \frac{6}{10}$

Ⓖ $\frac{1}{4} > \frac{1}{2}$

Ⓗ $\frac{1}{2} > \frac{4}{10}$

Ⓙ $\frac{1}{10} > \frac{1}{4}$

3 What unit would be best to measure the width of a penny（美分硬币）?

Ⓐ mile（英里）

Ⓑ yard（码）

Ⓒ meter（米）

Ⓓ millimeter（毫米）

4 4 x 10 = ______

Ⓕ 410

Ⓖ 14

Ⓗ 40

Ⓙ 6

5 What is the average width（平均宽度）of a classroom door?

Ⓐ 3.3 ft.

Ⓑ 3 m

Ⓒ 10.3 cm

Ⓓ 12.4 in.

6 Pick one question from this test. Explain how and why you chose your answer.

____

____

____

____

____

____

Perimeters of Shapes
图形的周长

Directions: Measure the perimeter（周长） in centimeters（厘米） of each of the shapes below. Round each measurement to the nearest 0.5 centimeter. Do you have to measure each side to work out the perimeter?

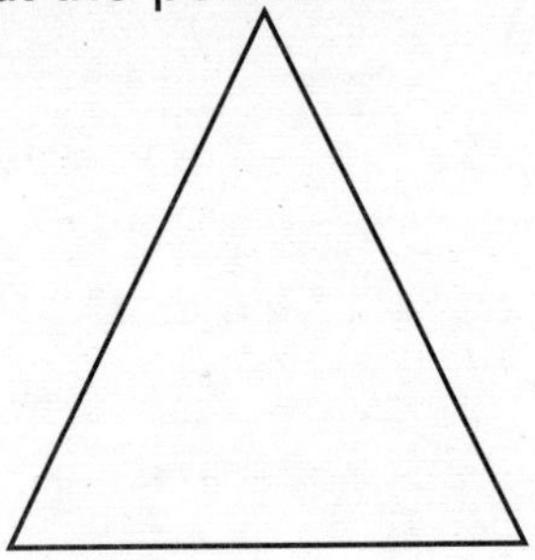

1. *P* = ____________________

2. *P* = ____________________

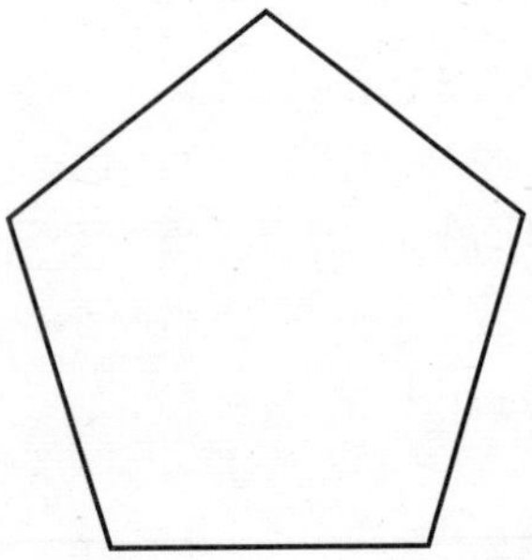

3. *P* = ____________________

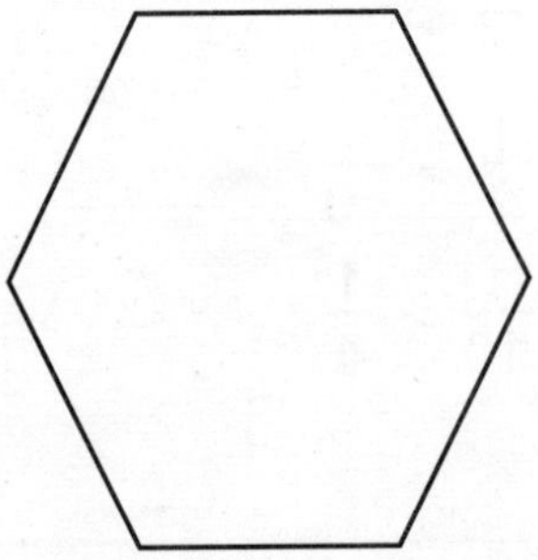

4. *P* = ____________________

Challenge

Directions: The perimeter of a rectangle（正方形） is 36 cm. What could the possible dimensions（尺寸） of the rectangle be? Give three examples.

1. ____________________ **2.** ____________________ **3.** ____________________

Car Trouble
小汽车的难题

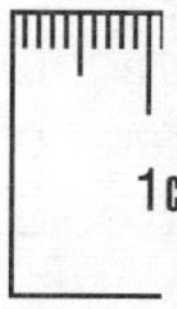

On John's ninth birthday, his dad gave him two collectible（收藏品）cars. John knew that they were special. They weren't really toys, so he wanted to display them on a shelf（架子）in his room. John worried about the cars. He knew his little sister didn't understand that they were just to look at. He worried she might play with them and accidentally（意外地）break the cars. John's mom said he should buy a display case（展示柜）for each of the cars.

"What's a display case?" asked John.

"A display case is a clear box that will keep the dust off（防止沾染灰尘）your collectibles. It will also keep your little sister from playing with them," explained his mother.

John thought this was a great idea. John decided he would need to know what size boxes to buy. John wasn't sure how big the cars were so he measured them with a ruler he found in his desk drawer（抽屉）.

Later that day John's mom took him to a hobby store（爱好品商店）to look for the perfect display cases. John had many choices. The display boxes were different lengths and heights. The man behind the counter（柜台）gave John a flyer（传单）with drawings of the actual boxes. The man said there were too many cases to get down, but he could choose the right size from the flyer.

Can you help John choose the right box for his new cars? Above are pictures of John's collectible cars.

Car Trouble *(cont.)*

What Is the Problem?

John wants to keep his collectible cars in good condition（条件）. He and his mother are going to the hobby store to find a display case for each of his cars. John measured his cars with a ruler, using centimeters（以厘米为单位）.

Use the information on page 73 and below to answer the questions. Before you begin solving each problem, be sure to locate the information you will need. Think about how you will create your table and what you will include in it.

Remember: If you have trouble planning ways to solve a problem, reread it and rethink your strategy. When you have a solution, think about whether or not your answer makes sense.

Problem-Solving Strategy: Creating a Table

Problem A

Before John went to the hobby store, he had to measure his cars. He decided to measure the smaller car first. Using the picture on the previous page（前一页）, make an estimate in centimeters of how long and how tall the car is. Use the centimeter ruler to find the actual length and height. Repeat the process for the larger car.

To help John at the store, organize the information in the table below. How close（接近的） were your estimates for both cars?

Object	Estimated Length 估计长度	Estimated Height 估计高度	Actual Length 实际长度	Actual Height 实际高度
Smaller Car				
Larger Car				

Problem B

John is trying to decide which display case to purchase（购买） for each car. Based on your actual measurements, what are the dimensions (in centimeters) of the smallest display case that John would be able to buy for the smaller car? What are the dimensions (in centimeters) of the smallest display case that John would be able to buy for the larger car?

Problem C

How much longer will the larger case need to be than the smaller case?

Extension

If John decides to put these cases end to end（首尾相连） on a shelf, how long (in centimeters) will the shelf need to be?

Standardized Test Preparation 17
标准考前测试 17

1 Look at the shaded part of each figure. Which figure shows $\frac{2}{3}$ shaded?

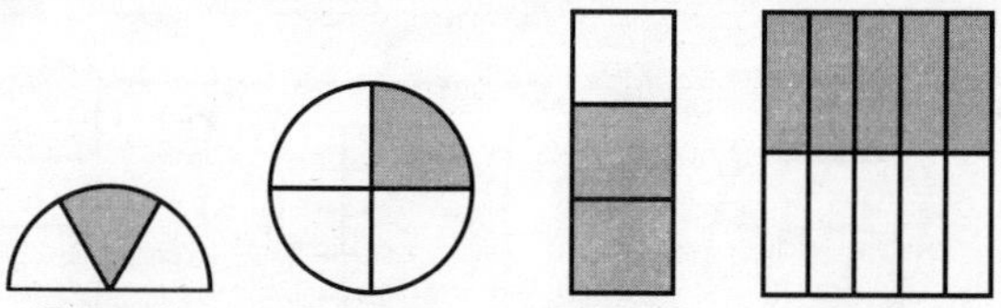

Figure 1 Figure 2 Figure 3 Figure 4

Ⓐ Figure 1

Ⓑ Figure 2

Ⓒ Figure 3

Ⓓ Figure 4

2 What is the next number in the pattern?

7, 14, 21, 28, ____

Ⓕ 47

Ⓖ 29

Ⓗ 37

Ⓙ 35

3 How many ounces of water are in this jug?

Ⓐ 1 oz.

Ⓑ 3 oz.

Ⓒ 2 oz.

Ⓓ 5 oz.

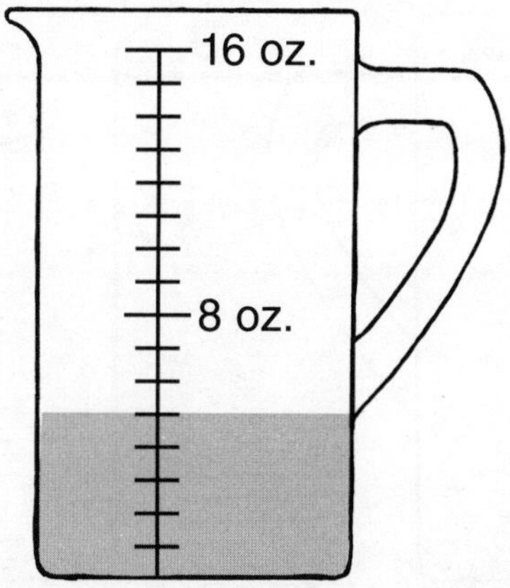

4 Round 252 to the nearest 10.

Ⓕ 260

Ⓖ 262

Ⓗ 250

Ⓙ 251

5 Which shows a picture graph（图形统计图）?

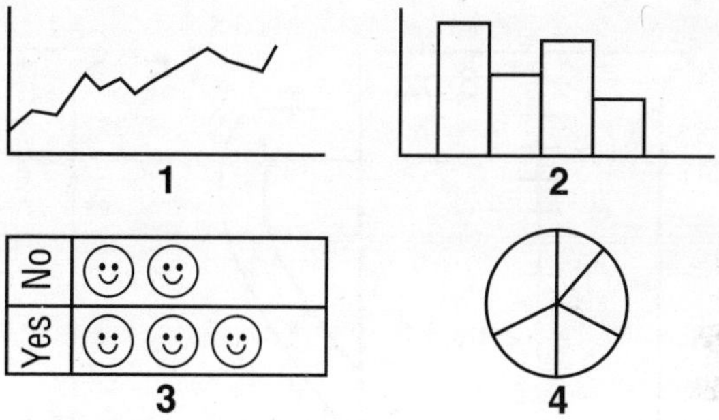

Ⓐ 1 Ⓒ 3

Ⓑ 2 Ⓓ 4

6 Pick one question from this test. Explain how and why you chose your answer.

____

____

____

____

____

____

Crazy Capacities

多变的容量

Directions: Mark the water level（水位）according to the number of ounces found below each jug.

1. 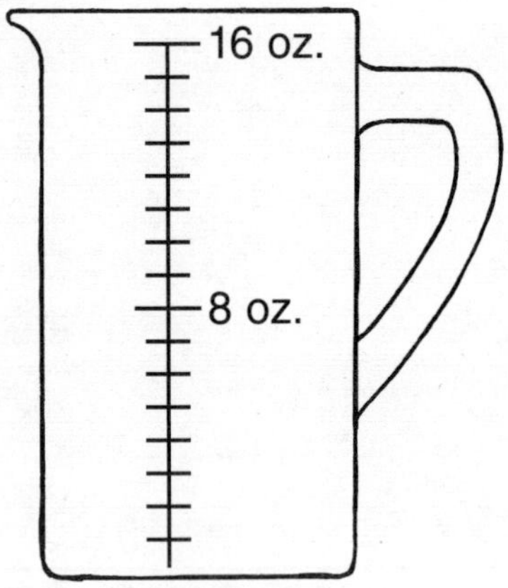

6 oz.

2. 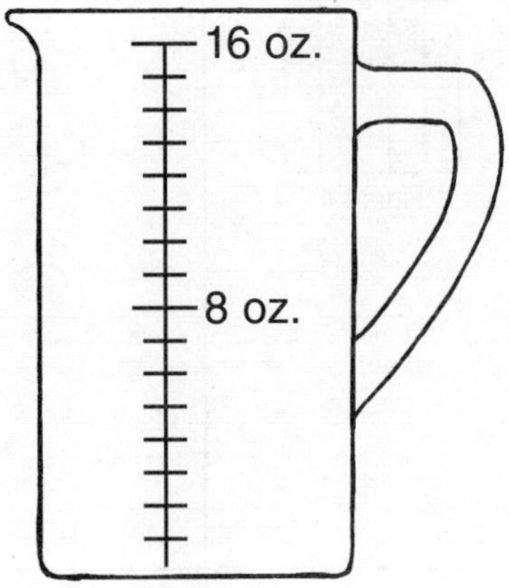

1 oz.

3. 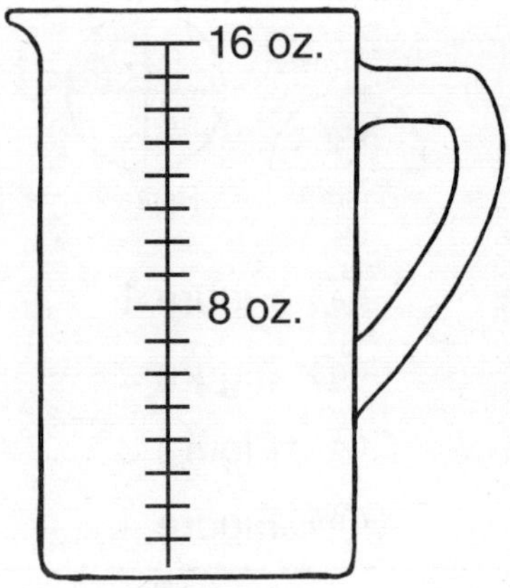

8 oz.

4. 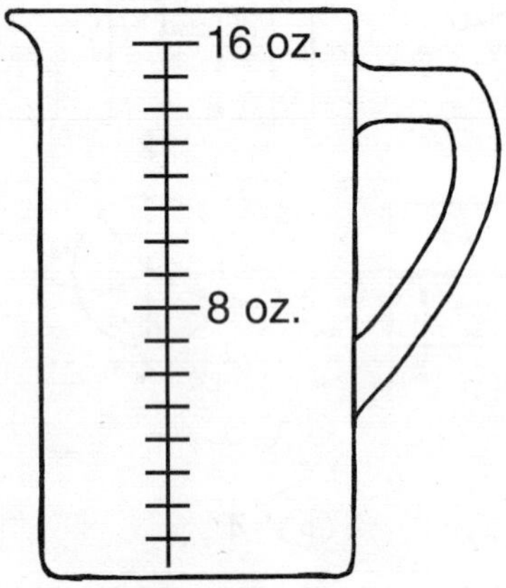

16 oz.

5. 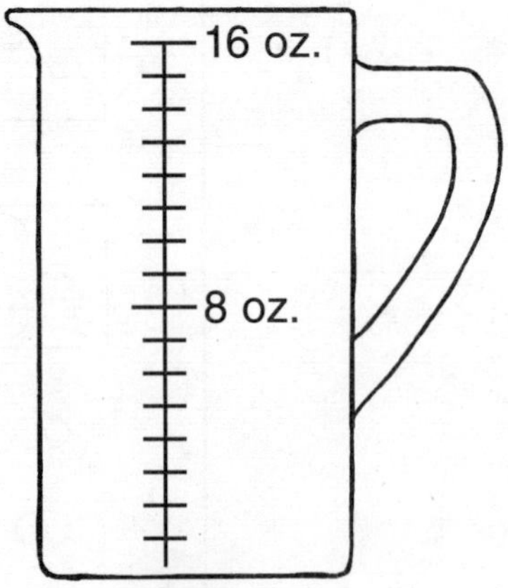

4 oz.

6. 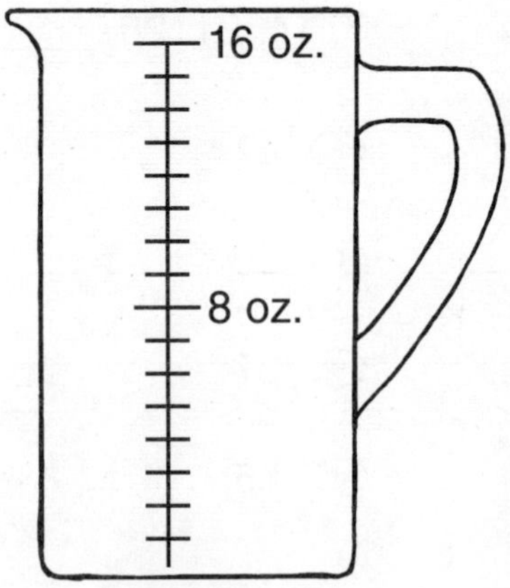

15 oz.

7. 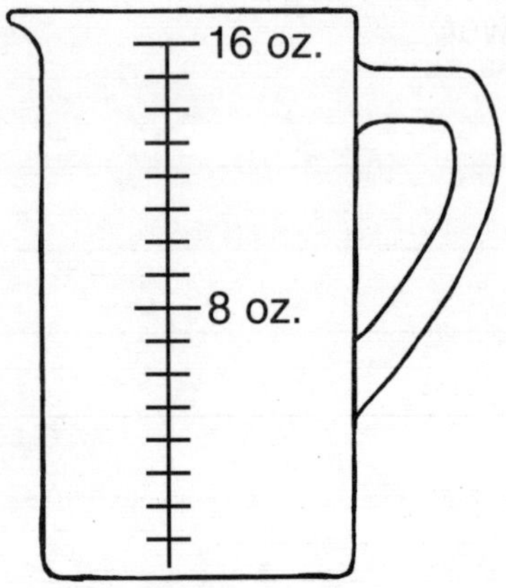

9 oz.

8. 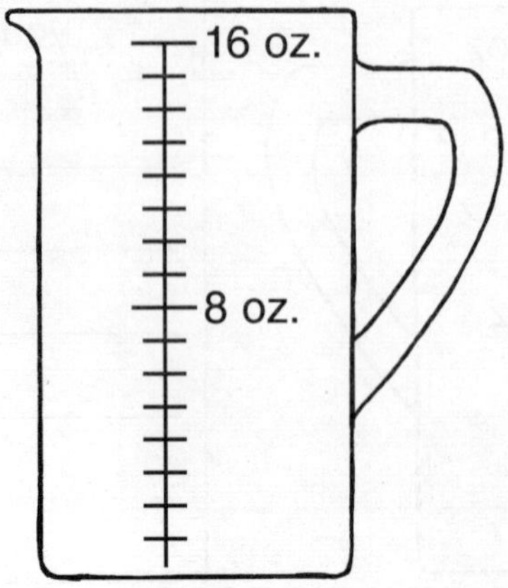

2 oz.

9. 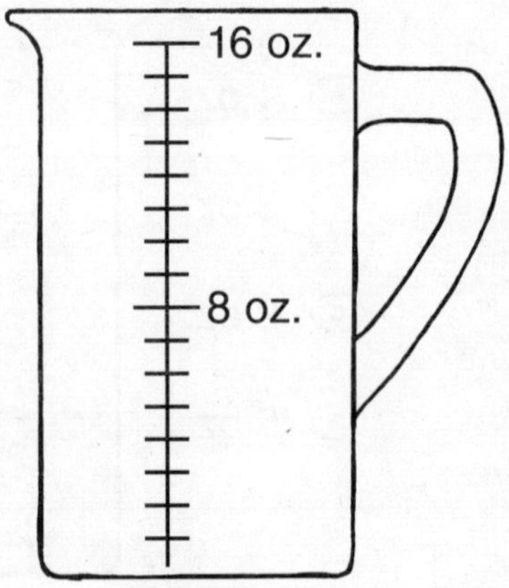

10 oz.

Measuring Capacities

容量的度量

Cups, pints, quarts, and gallons are used to measure liquids.（茶杯、品脱、夸脱和加仑常用于度量液体）

1 pint = 2 cups	1 quart = 4 cups	1 gallon = 16 cups

Directions: Answer each question.

1. About how much water could a mixing bowl（搅拌缸，沙拉碗）hold?

5 quarts　　5 pints

Ⓐ　　Ⓑ

2. About how many gallons would fill a sink（水槽，洗碗池）?

50 gallons　　5 gallons

Ⓐ　　Ⓑ

3. How much detergent（洗涤剂）should you use to wash clothes?

1 cup　　10 cups

Ⓐ　　Ⓑ

4. How many servings of punch（潘趣酒）are in one pitcher（大水罐）?

8 cups　　8 quarts

Ⓐ　　Ⓑ

5. How much milk could a baby drink at one time?

1 pint　　1 gallon

Ⓐ　　Ⓑ

6. How much juice should be bought for a family of 4?

1 pint　　1 quart

Ⓐ　　Ⓑ

7. How much coffee fills a coffee pot（咖啡壶）?

2 pints　　2 gallons

Ⓐ　　Ⓑ

8. How much water would you use to wash a car?

10 gallons　　1 gallon

Ⓐ　　Ⓑ

9. How much water would you use to brush your teeth?

1 gallon　　1 cup

Ⓐ　　Ⓑ

Creating a Table 2 Problems
制作表格 2 练习

Directions: Solve each problem and show your work.

Problem A

There are an equal number of chickens and sheep on a farm. There are 30 legs in all. How many chickens and sheep are there?

Problem B

Laura sells three strawberry pies（草莓派） for every four lemon pies（柠檬派）. If Laura sells 15 strawberry pies, how many lemon pies will she have sold? How many pies were sold in all?

Standardized Test Preparation 18

标准考前测试 18

1 What fraction of the shapes are circles?

Ⓐ $\frac{2}{10}$

Ⓑ $\frac{3}{10}$

Ⓒ $\frac{5}{10}$

Ⓓ $\frac{7}{10}$

2 Which number is not a multiple（倍数） of 3?

Ⓕ 23

Ⓖ 21

Ⓗ 30

Ⓙ 24

3 1 foot = _____

Ⓐ 6 in.

Ⓑ 12 in.

Ⓒ 25 cm

Ⓓ 100 cm

4 6 x 6 = _____

Ⓕ 0

Ⓖ 12

Ⓗ 36

Ⓙ 660

5 Which of these objects would hold（容纳） the most cups of liquid?

Ⓐ milk carton（牛奶盒）

Ⓑ water pail（水桶）

Ⓒ child's swimming pool（儿童游泳池）

Ⓓ bathroom sink（浴缸）

6 Pick one question from this test. Explain how and why you chose your answer.

____

____

____

____

____

____

Lesson 19

Estimating and Measuring Length
长度的估算与测量

Directions: Think about which measuring instrument（测量工具） would be most suitable to measure each object ［ruler, yardstick（码尺）, or tape measure（卷尺）］. Then, estimate each measurement. And finally, complete each measurement and record it in the chart.

Object	Measuring Instrument	Estimate	Measurement
length of board（木板）			
height of desk			
length of hall（门厅）			
width of hall			
length of playground			
length of pencil			
width of desk			
distance（距离） to lunchroom			

Parent Open House

家长开放日

A few minutes after 3:30 P.M., the door slammed shut（砰地关上）. Tommy was home from school. Tommy ran into the kitchen（厨房）to find his mom.

"Did you have a good day at school?" his mom asked.

"Oh yes! Guess what?" Tommy exclaimed.

"What?" asked his mom.

"Tomorrow is Parent Open House," said Tommy.

"What is that?" asked his mom.

"You can come to school with me and visit（参观）. You can even eat lunch with me," Tommy explained.

"That sounds great. What subjects（主题，科目）do you want me to visit?" asked his mom.

"You have to come to math, reading, music, and lunch," said Tommy.

"Wow! You have a lot of favorite（喜欢的）subjects. What time do I need to be there?"

"Well, music starts at 9:30. Reading is at 2:30. Math is at 11:30, right after P.E.（体育课）" Tommy said.

Tommy's mom said, "I know lunch is at 12:30. I guess I will be at your school for most of the day."

Parent Open House *(cont.)*

What Is the Problem?

Tommy's mom is going to the Parent Open House at his school. Tommy is excited. She is going to visit him on Tuesday. She needs to know Tommy's schedule（日程安排）if she is going to visit his favorite subjects. Each subject is one hour long.

Use the information in the problems below to create a table and answer the questions.

Before you begin solving each problem, think about what information is given and what information you will need to figure out to complete the problem. Think about how you will create your table and what you will include in it.

Remember: If you have trouble planning ways to solve each problem, reread it and rethink your strategy. When you have a solution, think about whether your answer makes sense.

Problem-Solving Strategy: Creating a Table

Problem A

Create a table of Tommy's day at school. School starts at 8:30 A.M. How much time is there between the start of school and math class?

Hint: Each class is one hour long.

Problem B

Tommy's mom dropped him off（让……下车）at school. She wants to run errands（工作，办事）. It will take her one hour to run all her errands. Will she be able to visit the first class Tommy wanted her to visit?

Problem C

Tommy's mom got to school at the start of music class. She stayed until school got out at 3:30 P.M. How long was Tommy's mom at school?

Hint: Add the time school ends to your table. Count the hours between music class and the end of school.

Extension

Tommy did not ask his mother to visit during science, P.E., or history. History class happens right after lunch. Add these subjects to his schedule on your existing table. Using all of the information you have gathered from both pages, at what time does Tommy go to science? At what time does he go to history?

Standardized Test Preparation 19
标准考前测试 19

1 What is the perimeter of this rectangle?

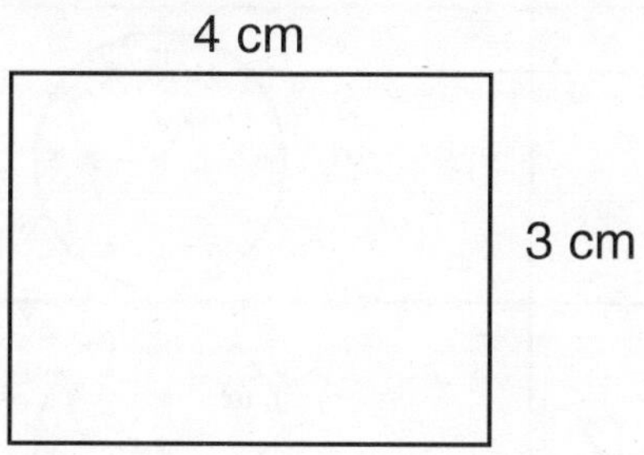

Ⓐ 14 cm
Ⓒ 11 cm
Ⓑ 12 cm
Ⓓ 7 cm

4 What is the next number in the pattern?

63, 54, 45, ____

Ⓕ 24
Ⓖ 27
Ⓗ 36
Ⓙ 34

2 _____ x 6 = 24

Ⓕ 3
Ⓖ 2
Ⓗ 4
Ⓙ 5

5 Choose the best category（范畴）for this statement: You will ride a camel（骆驼）during lunchtime.

Ⓐ Unlikely（不太可能）
Ⓑ Likely（可能）
Ⓒ Certain（肯定）
Ⓓ Impossible（绝不可能）

3 Which fraction is equal to $\frac{4}{5}$?

Ⓐ $\frac{8}{10}$
Ⓒ $\frac{1}{4}$
Ⓑ $\frac{2}{5}$
Ⓓ $\frac{4}{10}$

6 Pick one question from this test. Explain how and why you chose your answer.

____
____
____
____
____
____

Lesson 20

Digital and Analog
数字的和模拟的

Directions: Complete the chart.

Digital	Analog	Clock Face

	10 minutes to 7	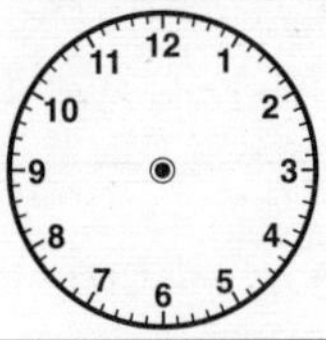
12:45		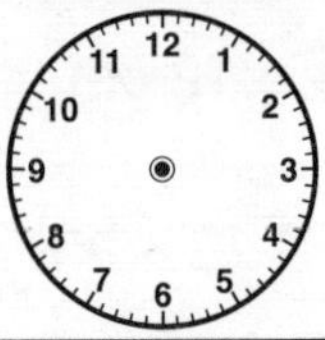
5:55		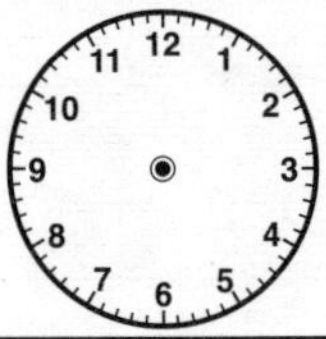
	11 minutes after 3	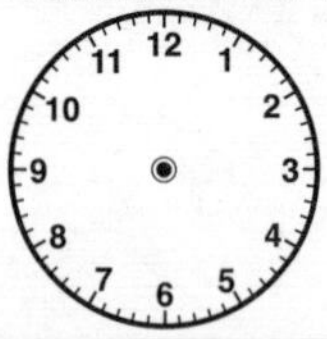
4:49		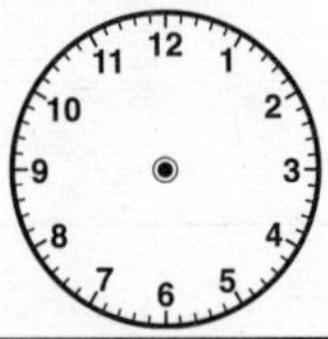
	26 minutes to 7	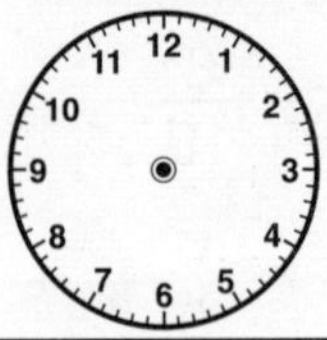

Challenge

Directions: Write pairs of digital times showing a difference of 30 minutes, so that the hour numbers are different. For example, 11:35 and 12:05 are 30 minutes apart and show different hours.

Matching Times
时间对对碰

Directions: Cut out（剪下）the cards along the solid lines（实线）. Match all of the cards together, showing three different representations of time.

	20 minutes to 11		9:36
quarter（一刻钟）after 2	4:55	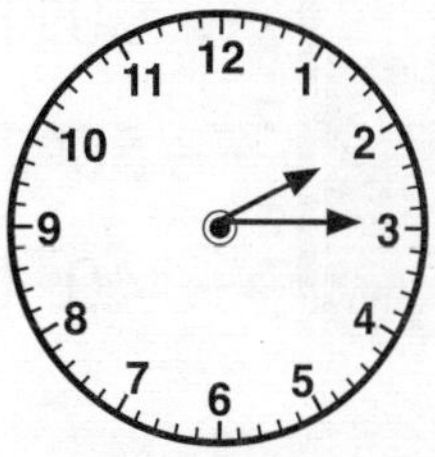	noon
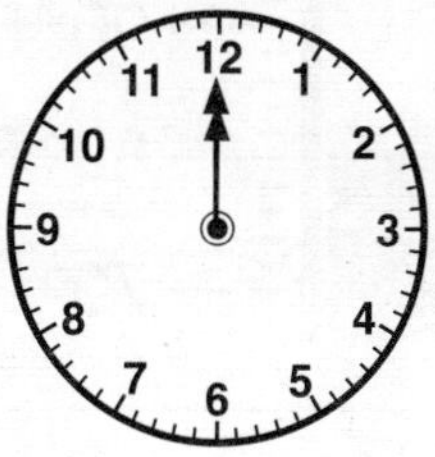	24 minutes to 10	5 minutes to 5	11:17
1:45	17 minutes after 11	12:00	
	6:05		quarter to 2
10:40	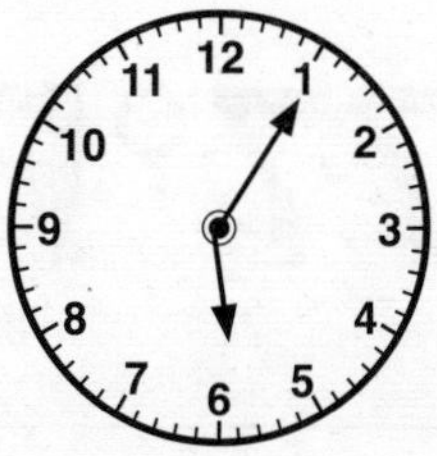	5 minutes after 6	2:15

Matching Temperatures
温度对对碰

Directions: Cut out the cards along the solid lines. Match each thermometer with the correct temperature.

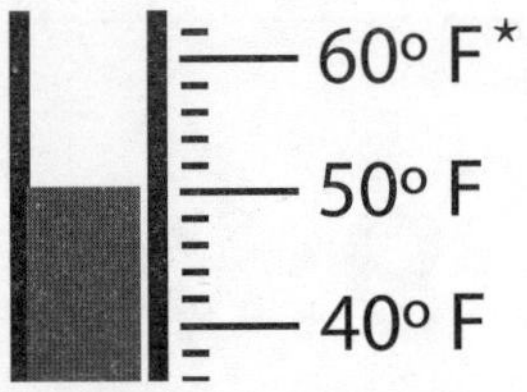

36°F

85°F

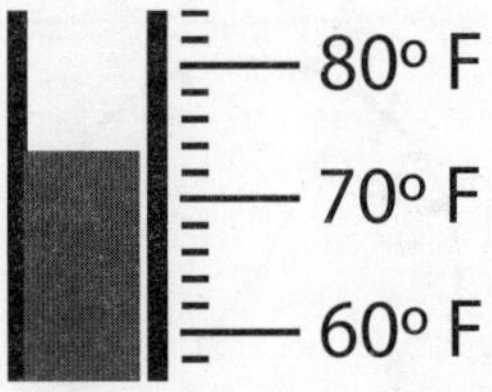

32°F

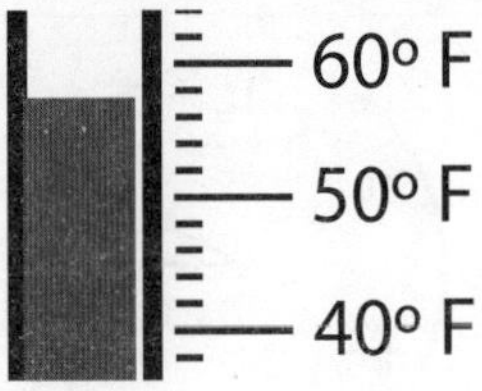

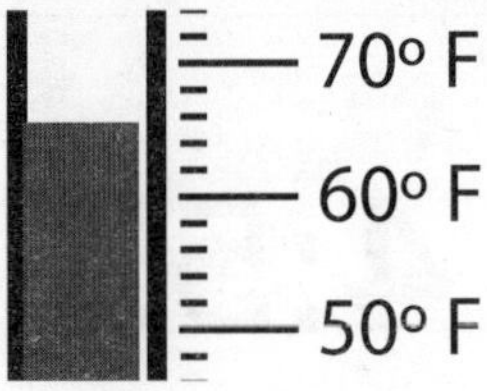

45°F

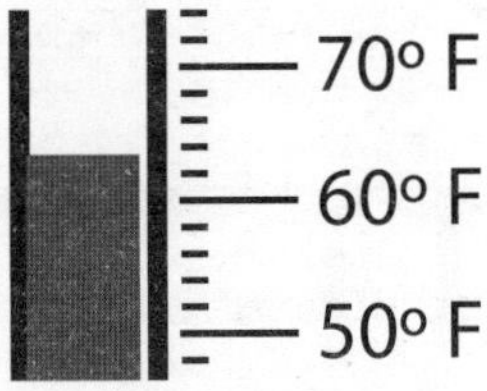

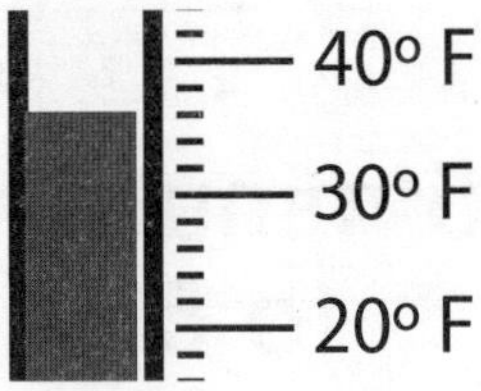

50°F

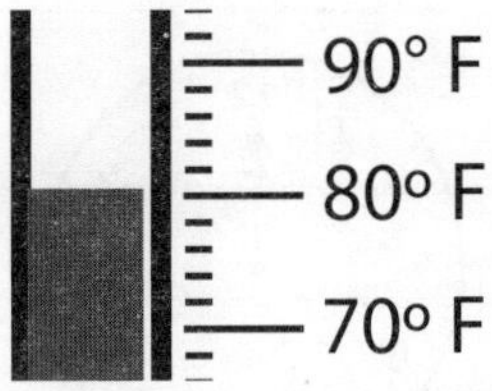

72°F

57°F

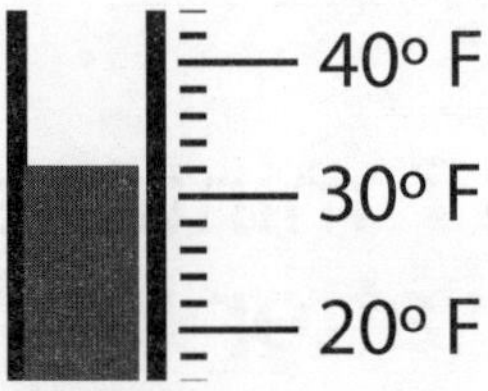

98°F

65°F

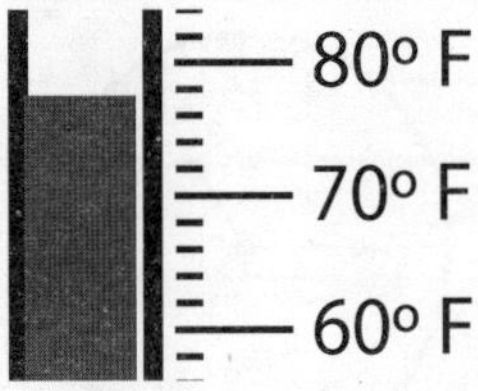

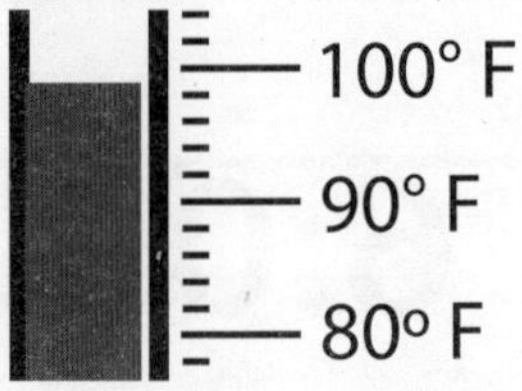

80°F

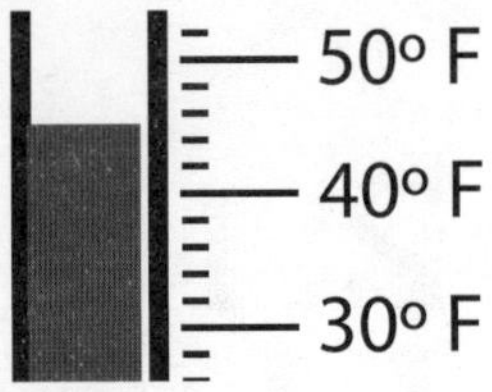

63°F

77°F

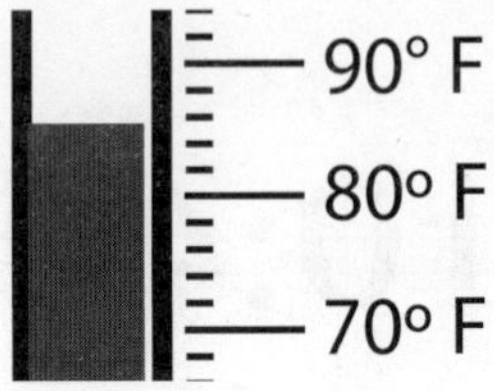

★华氏温度（℉）=32+摄氏温度（℃）×1.8。

Creating a Table 2 Group Problems

制作表格 2 分组练习

Directions: Solve Problem A and show your work. Using the strategy you've studied this week, write and solve your own problems for Problem B and Problem C.

Problem A

Anna earns 10 cents for each paper（报纸） she delivers（投递）. She puts 3 cents per paper in her piggy bank and spends the rest. If Anna delivers 10 papers, how much money does she earn, save, and spend?

Problem B: ______________________

____

____

____

____

____

Show work:

Answer: ______________________

Problem C: ______________________

____

____

____

____

____

Show work:

Answer: ______________________

Standardized Test Preparation 20
标准考前测试 20

1 Nate has 2 cats, 3 dogs, and 5 rabbits. What fraction of his pets are rabbits?

Ⓐ $\frac{7}{10}$

Ⓑ $\frac{5}{10}$

Ⓒ $\frac{3}{10}$

Ⓓ $\frac{2}{10}$

2 The double（两倍） of 24 is _____.

Ⓕ 6

Ⓖ 12

Ⓗ 48

Ⓙ 240

3 Which fraction is equivalent（相等的） to $\frac{6}{8}$?

Ⓐ $\frac{5}{8}$

Ⓑ $\frac{3}{4}$

Ⓒ $\frac{1}{2}$

Ⓓ $\frac{2}{3}$

4 About how tall is the average desk?

Ⓕ 2 ft.

Ⓖ 2 km

Ⓗ 5 cm

Ⓙ 5 in.

5 About 118 people attended（参加） the fair on _____.

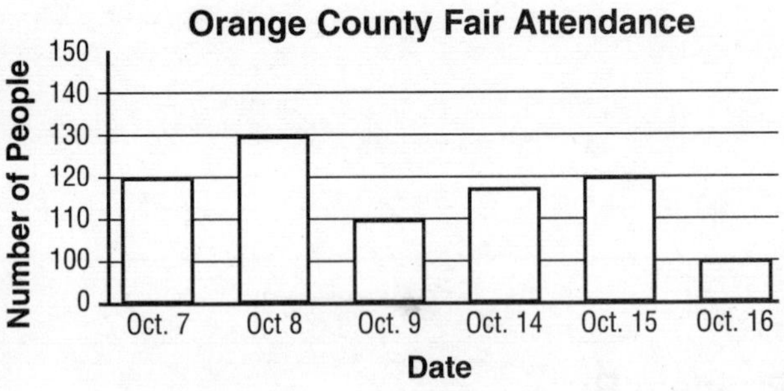

Ⓐ Oct. 8

Ⓑ Oct. 9

Ⓒ Oct. 14

Ⓓ Oct. 15

6 Pick one question from this test. Explain how and why you chose your answer.

____

____

____

____

____

____

Sorting Shapes

图形分类

Directions: Draw and write the names of at least two shapes that fit the description in each box.

1. These shapes have more than one line of symmetry (对称).	**2.** These shapes have all their sides the same length.
3. These shapes have straight (笔直的) sides.	**4.** These shapes have an even number (偶数) of sides.
5. These shapes have an odd number (奇数) of sides.	**6.** These shapes have an odd number of corners.

Acting It Out or Using Concrete Materials
演示或用小玩具解题

Using objects, such as counters（计数器，游戏筹码）or blocks（积木）, to represent（代表）people, places, or things in a problem often helps you find the answer. These objects can be moved through the steps of the problem and can help you keep track of what is happening. It can also be helpful to act out the roles of the different people or objects in the problem.

Using Concrete Materials

Some problems use large numbers. Others are best understood with objects that represent the real things in the problem.

Problem: Paper Squares

The Problem

Mr. Peters has four equal paper squares. He asks the class to join them so that they are connected by at least one edge（至少一条边相连）. How many different designs（设计）can be made?

Understanding the Problem

- *What do we know?*
  The four squares can be joined together in different ways, as long as they are connected by at least one edge.
- *What do we need to find out?*
  We need to find out the various ways that the squares can be joined together.

Planning and Communicating a Solution

The best way to solve this problem is to use four equal sized paper squares. Move the squares around to see how many different designs you can make. For each solution, draw a sketch of what the square combinations（组合）look like.

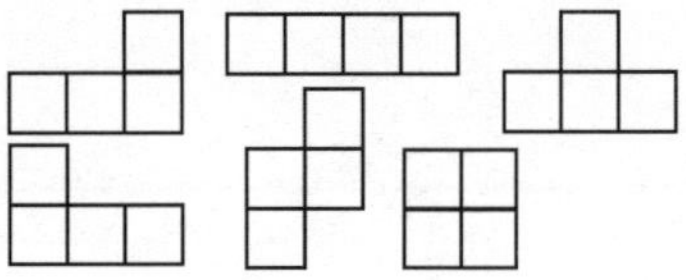

- *Do you see the answer?*
  You can connect the paper squares 6 different ways.

Acting It Out!

When a problem has things being moved around, acting out the problem can make it easier to find the solution.

Problem: Jellybeans

The Problem

Each time a student did something good, Mrs. Turner put one jellybean（豆形糖果）in a jar（罐）. If a student did something bad, she took one jellybean out of the jar.

The jar had eight jellybeans to start. Mrs. Turner took out three. Then she put in two. Later, she put in four. How many jellybeans are in the jar?

Understanding the Problem

- *What do we know?*
  Mrs. Turner is taking out jellybeans and putting in jellybeans.
- *What do we need to find out?*
  How many jellybeans are in the jar at the end of the day?

Planning and Communicating a Solution

Use real jellybeans (or counters) and a jar to actually act out the steps in this problem.

Follow the problem step-by-step（逐步地）, adding and taking away beans. Then count how many jellybeans you have left in the jar.

- *Do you see the answer?*
  Eleven jellybeans are in the jar.

Standardized Test Preparation 21

标准考前测试 21

1 Which fraction is less than $\frac{1}{2}$?

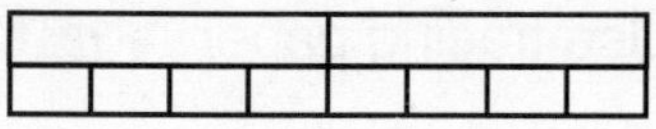

Ⓐ $\frac{5}{8}$

Ⓑ $\frac{3}{8}$

Ⓒ $\frac{4}{8}$

Ⓓ $\frac{6}{8}$

2 Which activity（活动） could you complete in 1 minute?

Ⓕ walking a mile

Ⓖ walking downstairs（下楼）

Ⓗ taking a bath（洗澡）

Ⓙ playing a basketball game

3 Choose the best category for this statement: You will travel to Venus（金星） tomorrow.

Ⓐ Certain

Ⓑ Likely

Ⓒ Unlikely

Ⓓ Impossible

4 Which produces the same answer as 9 x 5?

Ⓕ 5 + 9

Ⓖ 5 x 9

Ⓗ 9 – 5

Ⓙ 9 ÷ 5

5 Which of these shapes cannot have more than 3 corners?

Ⓐ square（正方形）

Ⓑ triangle（三角形）

Ⓒ rectangle（长方形）

Ⓓ octagon（八边形）

6 Pick one question from this test. Explain how and why you chose your answer.

Lesson 22

Congruent Rectangles and Triangles

全等四边形和全等三角形

Directions: Find the pairs of congruent shapes（全等图形）. Use tracing paper（描图纸）, if needed.

1. Draw a line connecting each pair of congruent rectangles（画线连接每对全等四边形）.

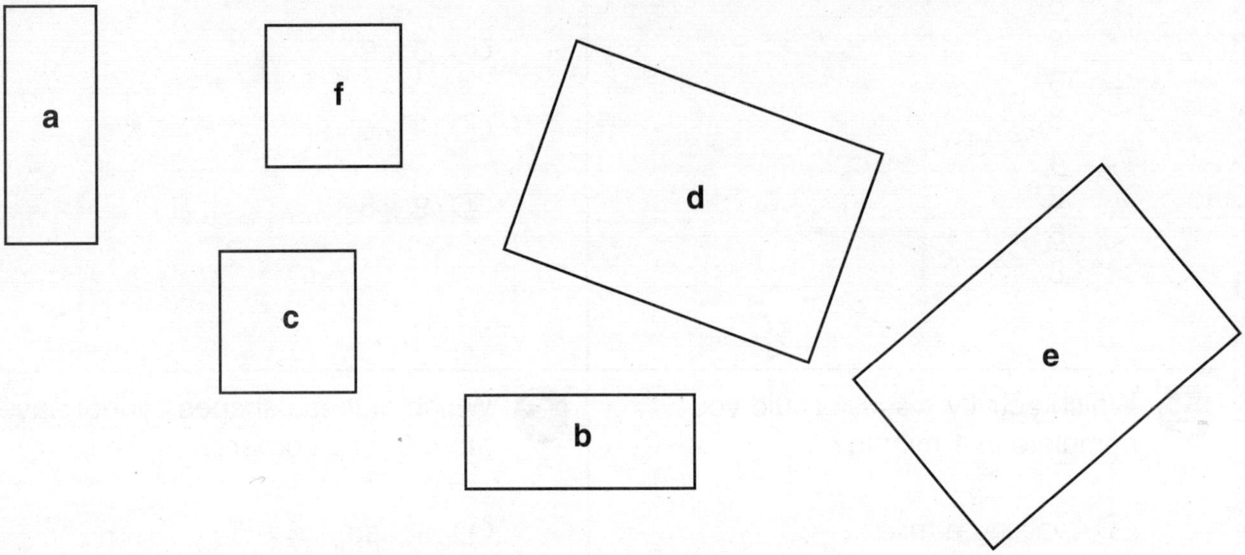

2. Draw a line connecting each pair of congruent triangles.

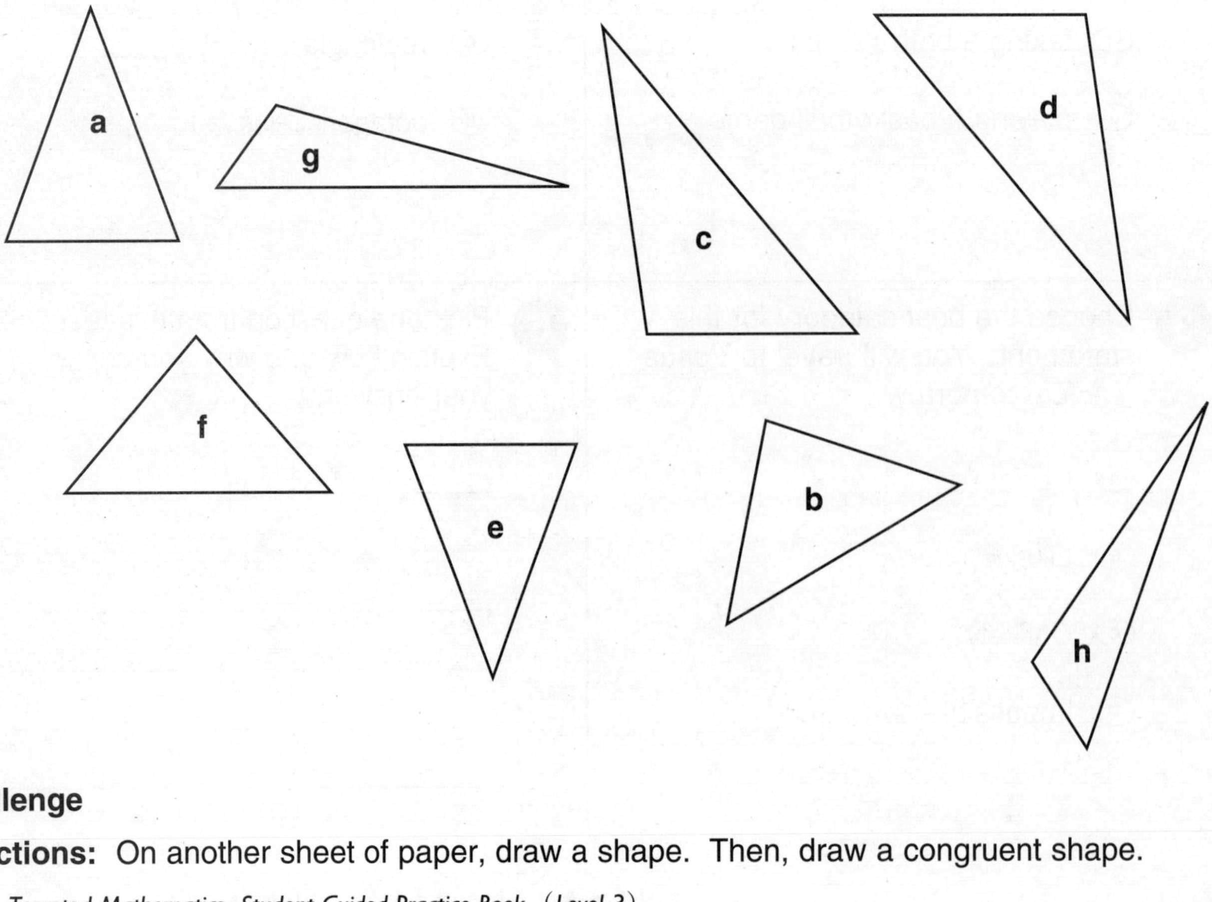

Challenge

Directions: On another sheet of paper, draw a shape. Then, draw a congruent shape.

Quilting Bee
大家缝聚会

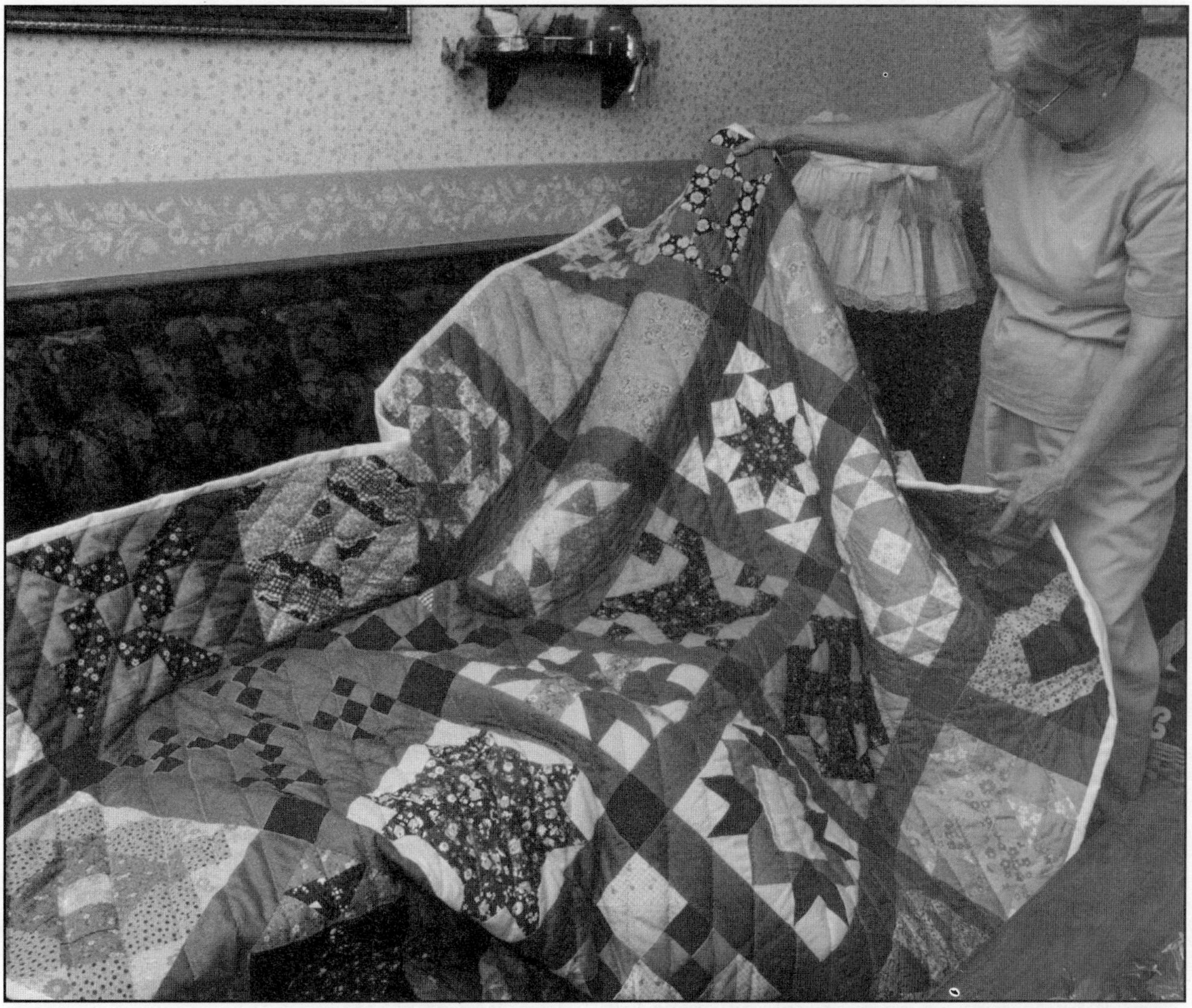

What did pioneer women do with scraps of material（碎布片）? They sewed（缝） them together into quilts（被子） to help keep their families warm. A quilt is a large blanket（毯子）. There are many different types of quilts.

The squares in the quilt shown above are made of fabric（布, 织物）, which is cut into shapes, such as squares, triangles, and rectangles. The shapes are pieced together to form patterns. In addition to beautiful geometric patterns（几何图案）, many quilt squares have symmetry（对称）. Pioneer women often got together to work on quilts at quilting bees. Women would either bring their own quilts, or they would work collectively（共同地） on one large quilt.

Look at the quilt above. Can you identify（辨认） the shapes that make up each square? Do the quilt squares have symmetry?

Quilting Bee *(cont.)*

What Is the Problem?

Your class has decided to design and assemble（组织） a paper quilt. Each student will design one square to be used in forming the larger quilt. Your job is to design a quilt square that will be used in the quilt. The measurements（尺寸） of the square must be eight inches by eight inches. Use paper cutouts（用纸剪下的图样） of shapes as concrete（具体的） objects to help you create your quilt square.

Use the information above and in the problems below to answer the questions. Before you begin solving each problem, think about what information is given and what information you will need to figure out to complete the problem. Think about how you will create your table and what information you will include in it.

Remember: If you have trouble planning ways to solve a problem, reread it and rethink your strategy. When you have a solution, think about whether your answer makes sense.

Strategy: Acting It Out or Using Concrete Materials

Problem A

Create a quilt square using at least two different two-dimensional shapes（平面图形）.

Problem B

Create a quilt square using at least two different two-dimensional shapes. The quilt pattern must be symmetrical from left to right.

Hint: Create the left side of the quilt square first. Then, use the left side of your quilt square to help you create the right side.

Problem C

Create a quilt square using at least two different two-dimensional shapes. The quilt pattern must be symmetrical from left to right and top to bottom（上下左右都对称）.

Hint: Create the top, left corner of the quilt square first. Then, use that corner to create the top, right corner. Finally, use the top half of the quilt to create the bottom half.

Extension

Use tape to put together your quilt squares. As a class, determine a layout（布局, 设计） that could be used to assemble all of the quilt squares. As a class, decide if you will need any "filler squares" to complete a row in order to maintain（保持） the shape of the quilt. How big will the finished quilt be?

Standardized Test Preparation 22
标准考前测试 22

1 The students in room 9 brought in 254 cans（罐头） for the canned food drive（捐赠食品罐头活动）. Mrs. Martinez brought in 14 more cans. How many total cans did room 9 bring in for the food drive?

Ⓐ 255 cans

Ⓑ 277 cans

Ⓒ 268 cans

Ⓓ 394 cans

2 What is the length of side x?

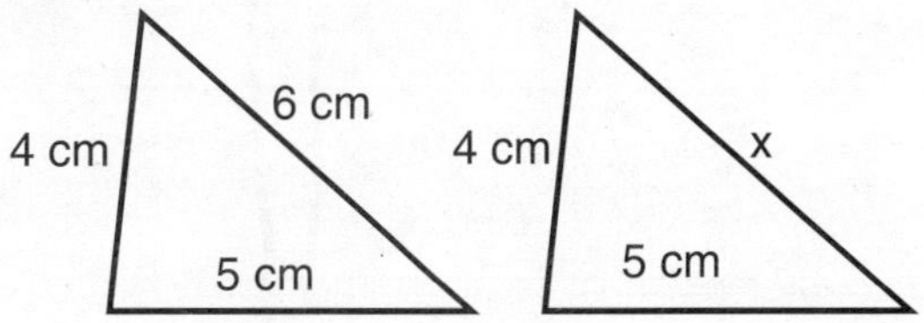

Ⓕ 6 cm

Ⓖ 15 cm

Ⓗ 4 cm

Ⓙ 5 cm

3 What is the perimeter of this rectangle?

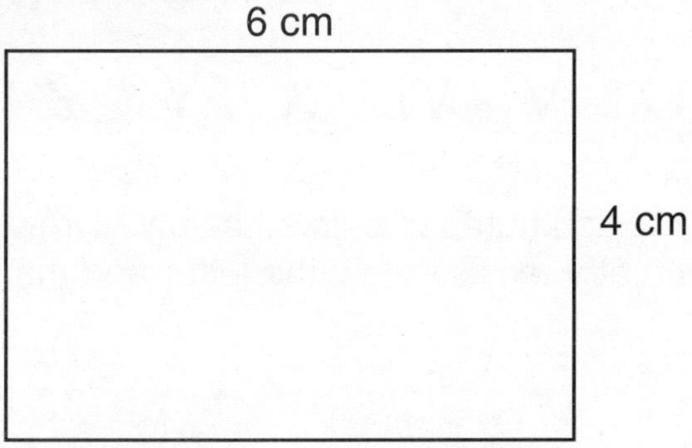

Ⓐ 6 cm　　Ⓒ 20 cm

Ⓑ 10 cm　　Ⓓ 24 cm

4 Which shows a bar graph（条形图）?

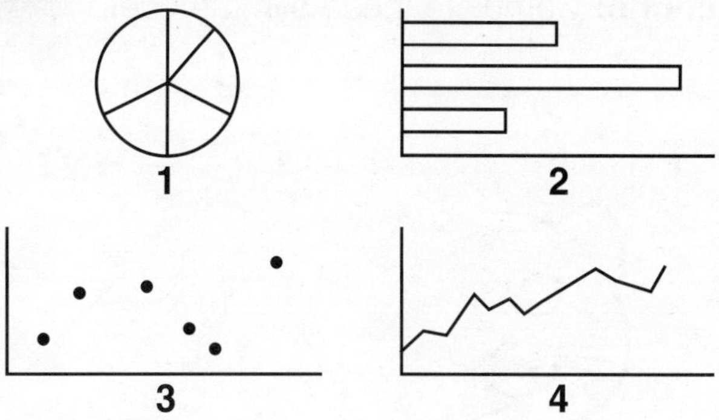

Ⓕ 1　　Ⓗ 3

Ⓖ 2　　Ⓙ 4

5 What rule has been used for the following group of numbers?

108, 208, 308, 408

Ⓐ subtract（减去） 10

Ⓑ multiply by（乘以） 10

Ⓒ add 100

Ⓓ subtract 8

6 Pick one question from this test. Explain how and why you chose your answer.

____

____

____

____

____

____

Lesson 23

Mirror Images

镜 像

Directions: Complete these symmetrical（对称的） reflections（映像） by drawing the missing parts.

1.

2.

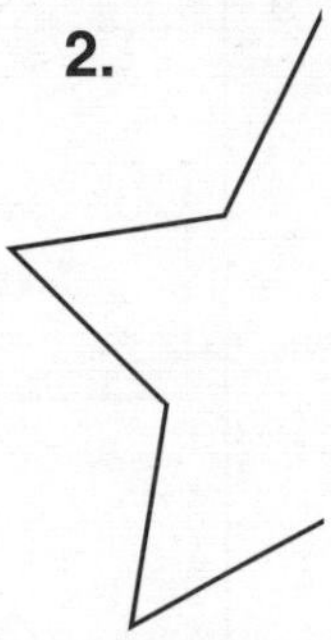

3.

4.

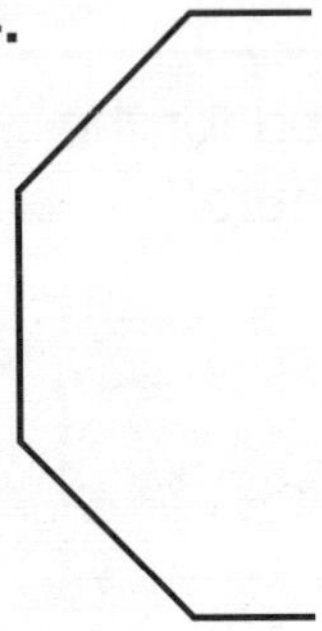

5.

Challenge

1. Which capital letters（大写字母） have reflective symmetry? Each capital letter is listed below. Draw the axis of symmetry（对称轴） for each one. If no axis of symmetry exists, circle the letter and mark an x through it.

A B C D E F G H I J K L M

N O P Q R S T U V W X Y Z

2. Which numbers (0–9) have reflective symmetry? Each number is listed below. Draw the axis of symmetry for each one. If no axis of symmetry exists, circle the letter and mark an x through it.

0 1 2 3 4

5 6 7 8 9

Acting It Out or Using Concrete Materials Problems
演示或用小玩具解题 练习

Directions: Solve each problem and show your work.

Problem A

Jasmine has an 8" (8英寸) square piece of paper. She cuts the paper in half. She takes one of the halves and cuts it in half. She does this three more times. How many pieces of paper does Jasmine now have?

Problem B

Jonah had a $20 bill (纸币). He exchanged the bill for two $10 bills. He took one of the $10 bills and exchanged (兑换) it for two $5 bills. He took one of the $5 bills and exchanged it for five $1 bills. Jonah took a $1 bill and exchanged it for four quarters (25美分硬币). What bills and coins (硬币) does Jonah now have?

Standardized Test Preparation 23
标准考前测试 23

1 How many 3rd graders are represented altogether in the pictograph（统计图表）?

3rd Graders' Favorite Colors

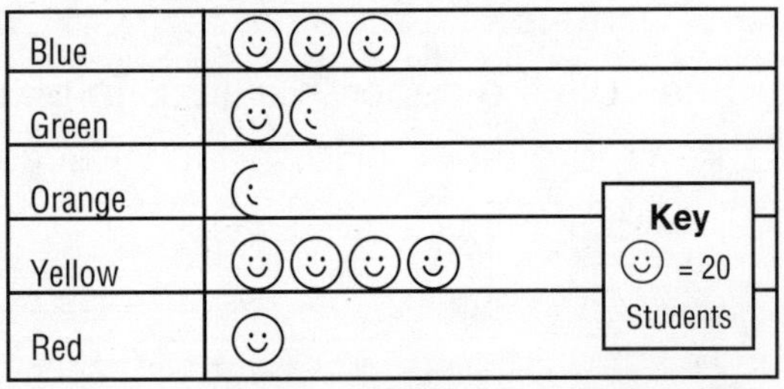

Ⓐ 10 students Ⓒ 130 students

Ⓑ 80 students Ⓓ 200 students

2 This chart shows the train schedule（列车时刻表）. Ruby is waiting for the train to Connecticut（美国康涅狄格州）. Right now, the time is 1:30 P.M. How long will Ruby wait for the next train to Connecticut?

Connecticut: 1:00 P.M., 2:15 P.M., 3:30 P.M.
New York: 5:30 P.M., 6:30 P.M., 7:00 P.M.
Boston: 7:00 P.M., 8:30 P.M., 10:00 P.M.

Ⓕ 30 minutes Ⓗ 2 hours

Ⓖ 45 minutes Ⓙ 3 hours

3 Round 125 to the nearest 10.

Ⓐ 100

Ⓑ 120

Ⓒ 130

Ⓓ 150

4 If the perimeter of a right triangle（直角三角形）is 24 cm long, how long is the perimeter of a congruent（全等的）right triangle?

Ⓕ 12 cm

Ⓖ 24 cm

Ⓗ 18 cm

Ⓙ 48 cm

5 On which two dates was the attendance（出席人数）the same?

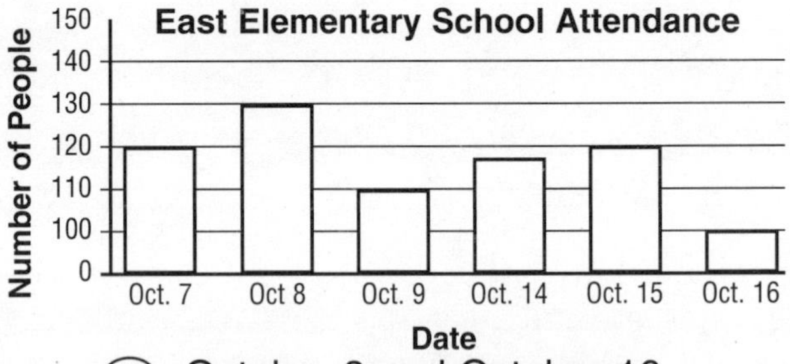

Ⓐ October 8 and October 16

Ⓑ October 7 and October 15

Ⓒ October 14 and October 16

Ⓓ October 9 and October 15

6 Pick one question from this test. Explain how and why you chose your answer.

____

____

____

____

____

____

Lesson 24

Creating Patterns

找规律

Directions: Continue each pattern to answer the question.

1. 

1. What is the 13th shape in this pattern?

2. 

2. What is the 15th shape in this pattern?

3. 

3. What is the 17th shape in this pattern?

4. 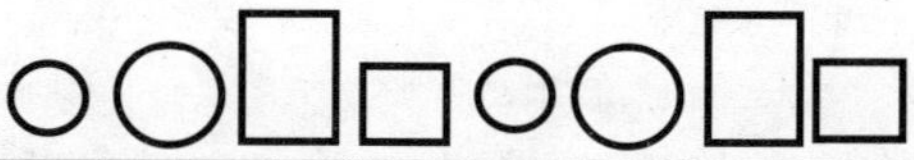

4. What is the 14th shape in this pattern?

5. 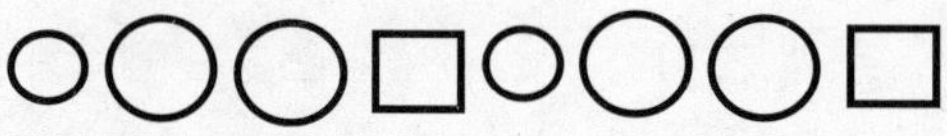

5. What is the 16th shape in this pattern?

6. 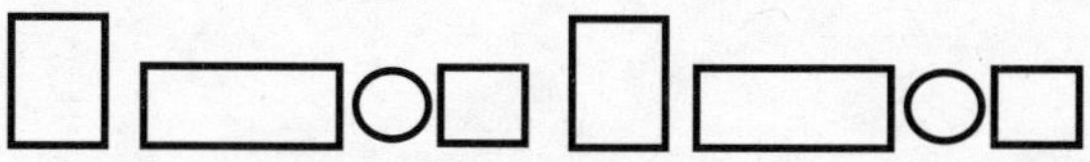

6. What is the 18th shape in this pattern?

Getting Around Town

在城里转转

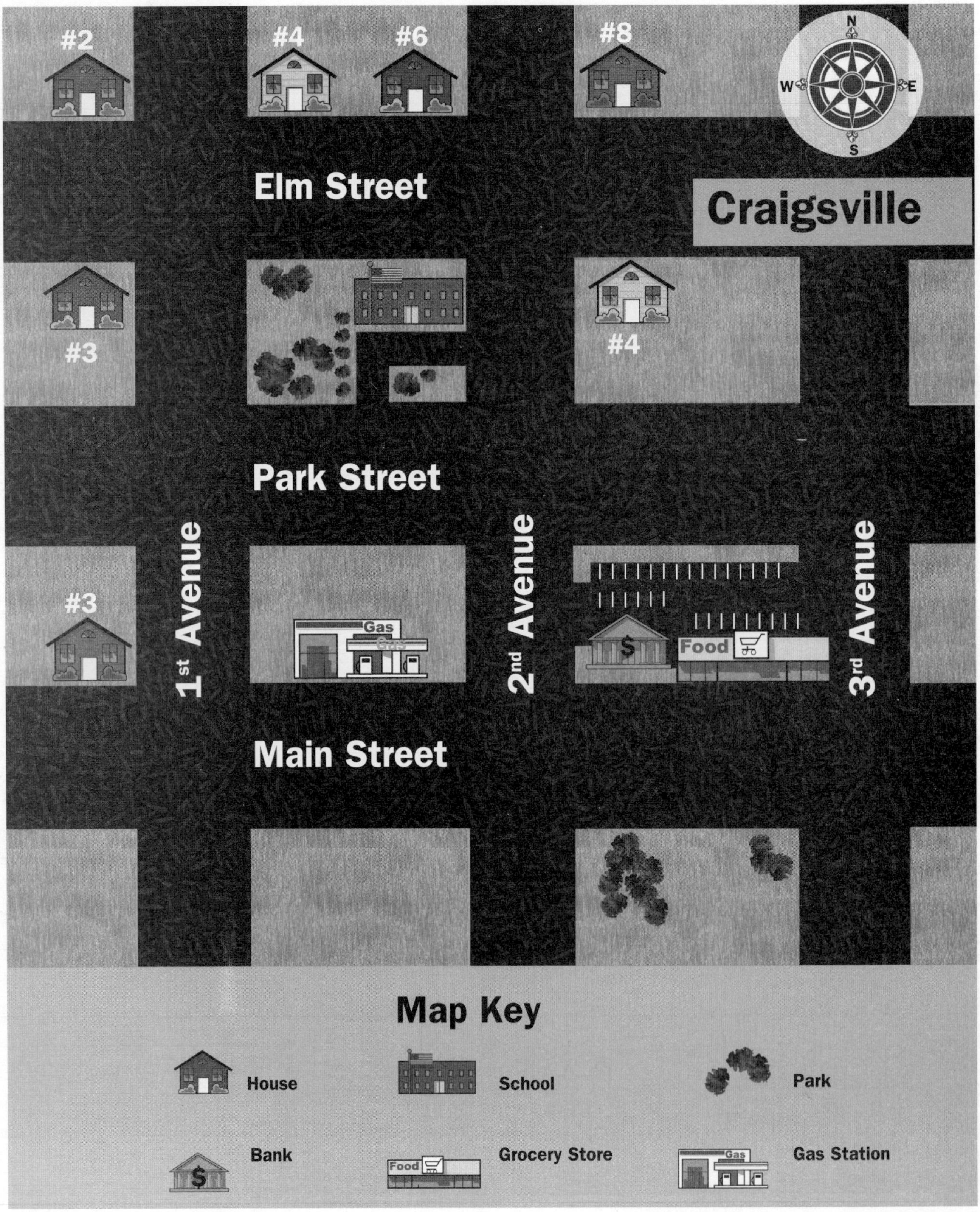

Getting Around Town *(cont.)*

What Is the Problem?

You live at 2 Elm Street in Craigsville. Your grandma is visiting you and does not know her way around town. Your job is to use the map to give directions（指路） and help her get around town. For example, if your grandma wanted to visit a friend at 8 Elm Street, she should go one block（街区） east.

Use the map on page 100 to answer the questions. Before you begin solving each problem, be sure to locate the information you will need. Think about how you will act out the problem or what concrete materials you will use to help you solve the problem.

Remember: If you have trouble planning ways to solve a problem, reread it and rethink your strategy. When you have a solution, think about whether your answer makes sense.

Strategy: Acting It Out or Using Concrete Materials

Problem A

Give compass directions to get from 2 Elm Street to the bank.

Hint: Use the Map Key to locate the bank.

Problem B

Give compass（指南针） directions for two different routes（路线） to get from 2 Elm Street to the grocery store（食品杂货店）.

Problem C

Your grandma wants to pick you up（中途停车接人） from school, go to the grocery store, and get gas（加油） before taking you home. Give compass directions for a route she can take.

Hint: Your grandma starts at your house.

Extension

Locate a map of your town. Plot（标绘） your school and a grocery store on the map. Give compass directions to tell how to get from the school to the grocery store.

Standardized Test Preparation 24
标准考前测试 24

1 Which is the best estimate of the time it would take for bushes（灌木）to grow from a planted seed（种下的种子）?

Ⓐ 30 seconds

Ⓑ 30 minutes

Ⓒ 30 hours

Ⓓ 30 days

2 What rule has been used for the following group of numbers?

95, 85, 75, 65

Ⓕ add 10

Ⓖ add 5

Ⓗ subtract 10

Ⓙ subtract 5

3 How many lines of symmetry can be drawn on this hexagon（六边形）?

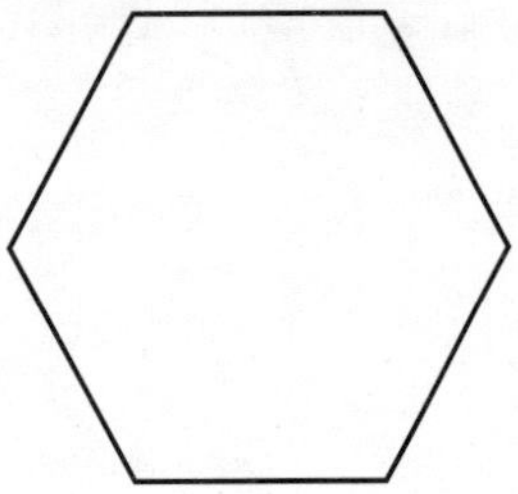

Ⓐ 1　　Ⓒ 3

Ⓑ 4　　Ⓓ 6

4 What word best describes these two figures?

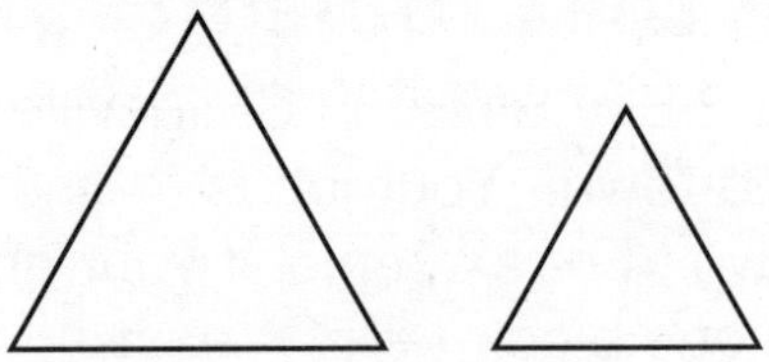

Ⓕ similar（相似的）

Ⓖ congruent（全等的）

Ⓗ right（直角的）

Ⓙ polyhedron（多面体）

5 Which of the following statements is certain?

Ⓐ Dinosaurs（恐龙）eat meat.

Ⓑ Everyone in your community eats meat.

Ⓒ Carnivores（食肉动物）eat meat.

Ⓓ Adults eat meat.

6 Pick one question from this test. Explain how and why you chose your answer.

____

____

____

____

____

____

Lesson 25

Three-Dimensional Shapes

3D（立体）图形

Directions: Complete the chart.

Shape	Name	Faces（面）	Edges（边）	Corners（角）
1. 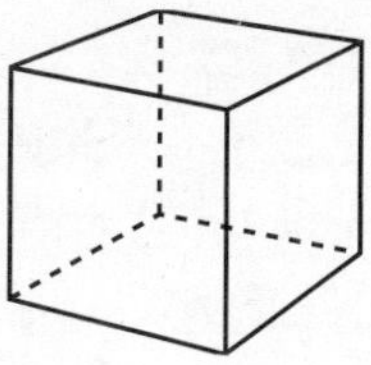				
2. 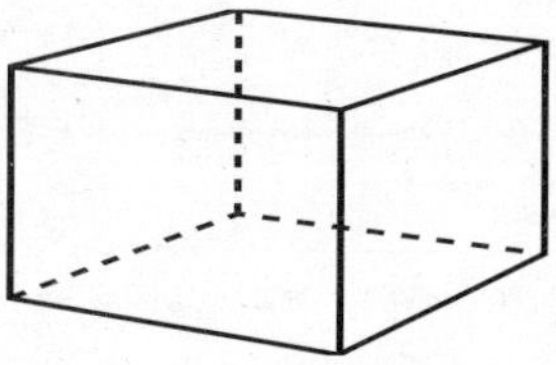				
3. 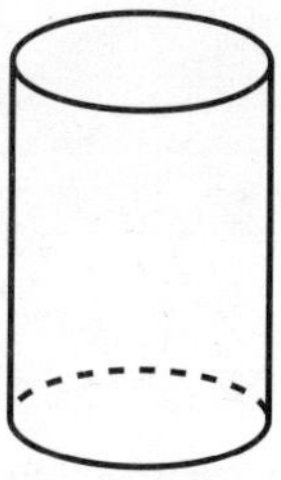				
4. 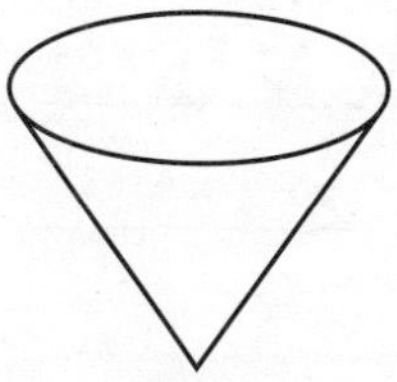				
5. 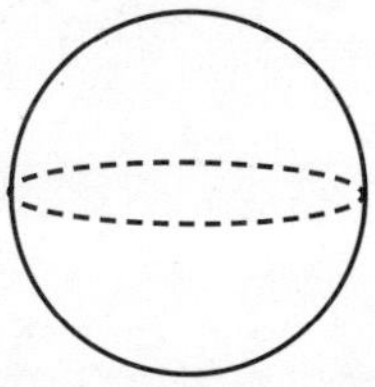				
6. 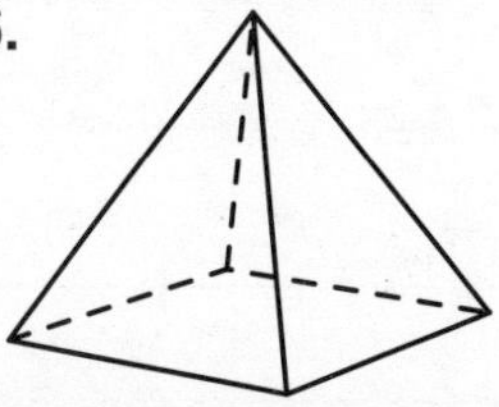				

Acting It Out or Using Concrete Materials Group Problems

演示或用小玩具解题 分组练习

Directions: Solve Problem A and show your work. Using the strategy you've studied this week, write and solve your own problems for Problem B and Problem C.

Problem A

Adrianna has a dozen（一打, 12个） cupcakes（杯形蛋糕） and three plates（盘子）. She puts half of the cupcakes on the first plate, two fewer cupcakes on the second plate, and the rest of the cupcakes on the third plate. How many cupcakes are on each plate?

Problem B: ______________________

____

____

____

____

____

Show work:

Answer: ______________________

Problem C: ______________________

____

____

____

____

____

Show work:

Answer: ______________________

Standardized Test Preparation 25

标准考前测试 25

1 What rule has been used for this group of numbers?

382, 392, 402, 412

Ⓐ add 100

Ⓑ multiply by 100

Ⓒ subtract 10

Ⓓ add 10

2 Which two watches（手表） show the same time?

Ⓕ

Ⓗ

Ⓖ

Ⓙ

3 Which two pieces of equipment received the same number of votes?

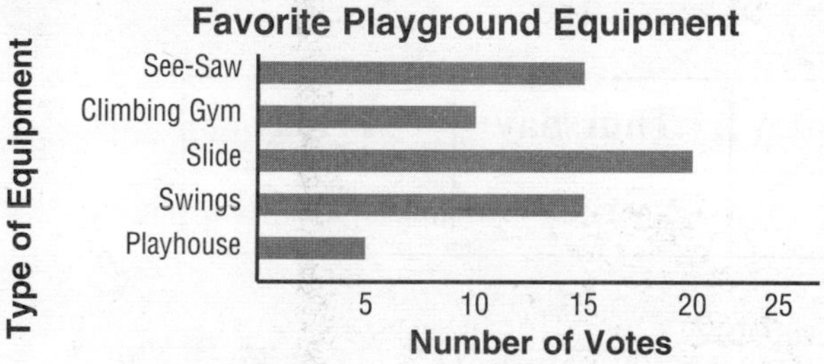

Ⓐ playhouse and climbing gym

Ⓑ see-saw and slide

Ⓒ climbing gym and swings

Ⓓ swings and see-saw

4 How many lines of symmetry can be drawn on this equilateral triangle?

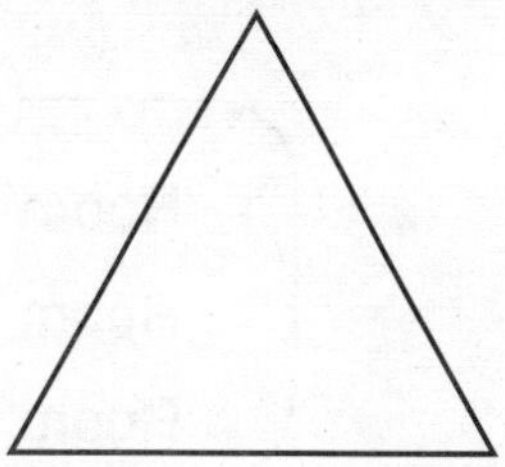

Ⓕ 1 Ⓗ 2

Ⓖ 3 Ⓙ 4

5 Which of the following statements is likely?

Ⓐ You will be captured（俘获） by aliens（外星人） today.

Ⓑ You will win a billion（十亿） dollars today.

Ⓒ You will see a flower somewhere today.

Ⓓ You will become a professional（职业的） football player this afternoon.

6 Pick one question from this test. Explain how and why you chose your answer.

____

____

____

____

____

____

Lesson 26

Raffle Tickets

彩 券

1. Make a picture graph that includes the following data:

Tickets Sold in a Week	
Room 1: 160	Room 4: 280
Room 2: 150	Room 5: 270
Room 3: 210	Room 6: 190

KEY
▭ = 20 tickets

____
(title)

Room 1	
Room 2	
Room 3	
Room 4	
Room 5	
Room 6	

2. The chart（表）below shows how many tickets students in Room 4 sold each day. Answer the following questions and then make a picture graph on another sheet of paper to show this information.

Monday	Tuesday	Wednesday	Thursday	Friday
60	50	90	33	47

a. On which day did Room 4 sell the most tickets? ______________

b. On which day was the least number of tickets sold? ______________

c. How many tickets did Room 4 students sell altogether? ______________

d. How many more tickets did they sell on Tuesday than on Friday? ______________

Drawing a Diagram
画图解题

Drawing a diagram（图表）or picture of a word problem（应用题）often helps you visualize（形象化）, or picture in your mind, the information in a problem more clearly.

You can draw a diagram using simple symbols（符号）or pictures. Drawing diagrams also helps you keep track of the information in problems that have more than one step. Diagrams allow you to organize your thoughts, which makes the problem simpler.

Before you begin using diagrams, read the following information to learn more about how and when to use diagrams to solve problems.

Problem: Cookies for School

The Problem

You are making cookies（小甜饼，曲奇）to take with you to school. You need to cut the cookie dough roll（揉成长条形的生面团）into 10 pieces. How many cuts do you need to make?

Understanding the Problem

- *What do I know?*
  I need to cut 10 pieces of cookie dough.
- *What do I need to find out?*
  How many cuts I will make.
- *Question to think about:*
  Is this a problem I can solve with simple math calculations, or should I draw a diagram?

Planning and Communicating a Solution

To solve this problem, draw a line to represent the cookie dough roll. Show the cuts you would need to make to get 10 cookies. Count or number the cuts as you draw them.

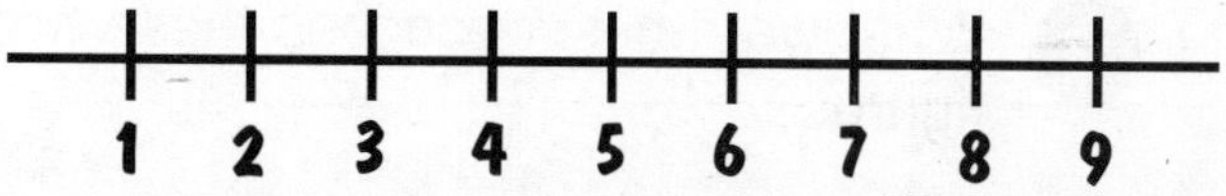

Do you see the answer?

To make 10 cookies, you need to cut the dough in 9 places.

Reflecting and Generalizing

To solve this problem, you could have simply used the information from the problem and thought "10 cookies, 10 cuts." But that solution would not be correct.

Extension

- If you make 9 cuts in each roll, how many rolls of dough would you need to get 80 cookies?
- How many cuts would you need to make if you had two rolls of dough and you wanted 22 cookies total?

Standardized Test Preparation 26

标准考前测试 26

1 How many lines of symmetry can be drawn on this triangle?

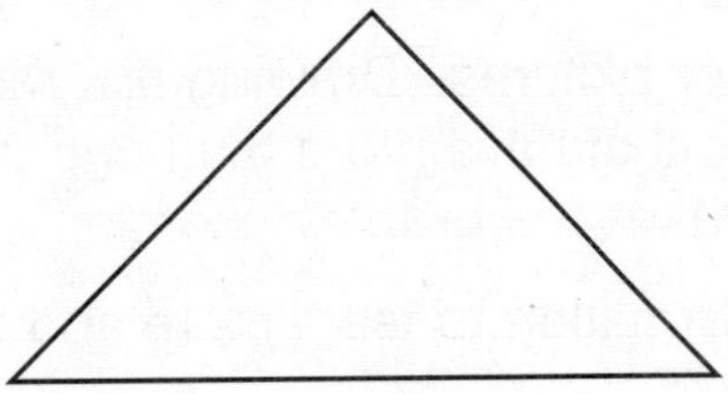

Ⓐ 0 Ⓒ 1

Ⓑ 2 Ⓓ 3

2 What word best describes these two figures?

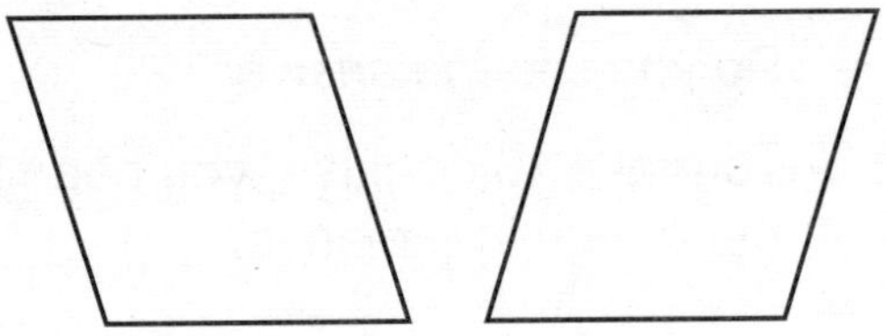

Ⓕ similar Ⓗ congruent

Ⓖ quadrilateral（四边形的） Ⓙ scalene（不等边的）

3 You arrived at school at 9:15. You left your house at 8:45. How long did it take you to get to school?

Ⓐ 30 minutes

Ⓑ 1 hour

Ⓒ 45 minutes

Ⓓ 15 minutes

4 What is the name of this shape?

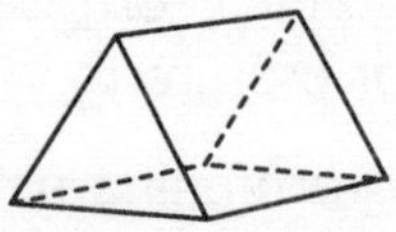

Ⓕ cube（立方体）

Ⓖ prism（棱柱）

Ⓗ pyramid（金字塔形）

Ⓙ sphere（球体）

5 Choose the best category for this statement: You will breathe（呼吸）air in the next five minutes.

Ⓐ Impossible

Ⓑ Unlikely

Ⓒ Likely

Ⓓ Certain

6 Pick one question from this test. Explain how and why you chose your answer.

____

____

____

____

____

____

Lesson 27

Our Heights
我们的身高

Directions: Use the bar graph（条形图）below to fill in height data for 10 of your classmates. Add a title（表头）and labels（标记）. Write the student names by the numbers below.

Title: ______________________________

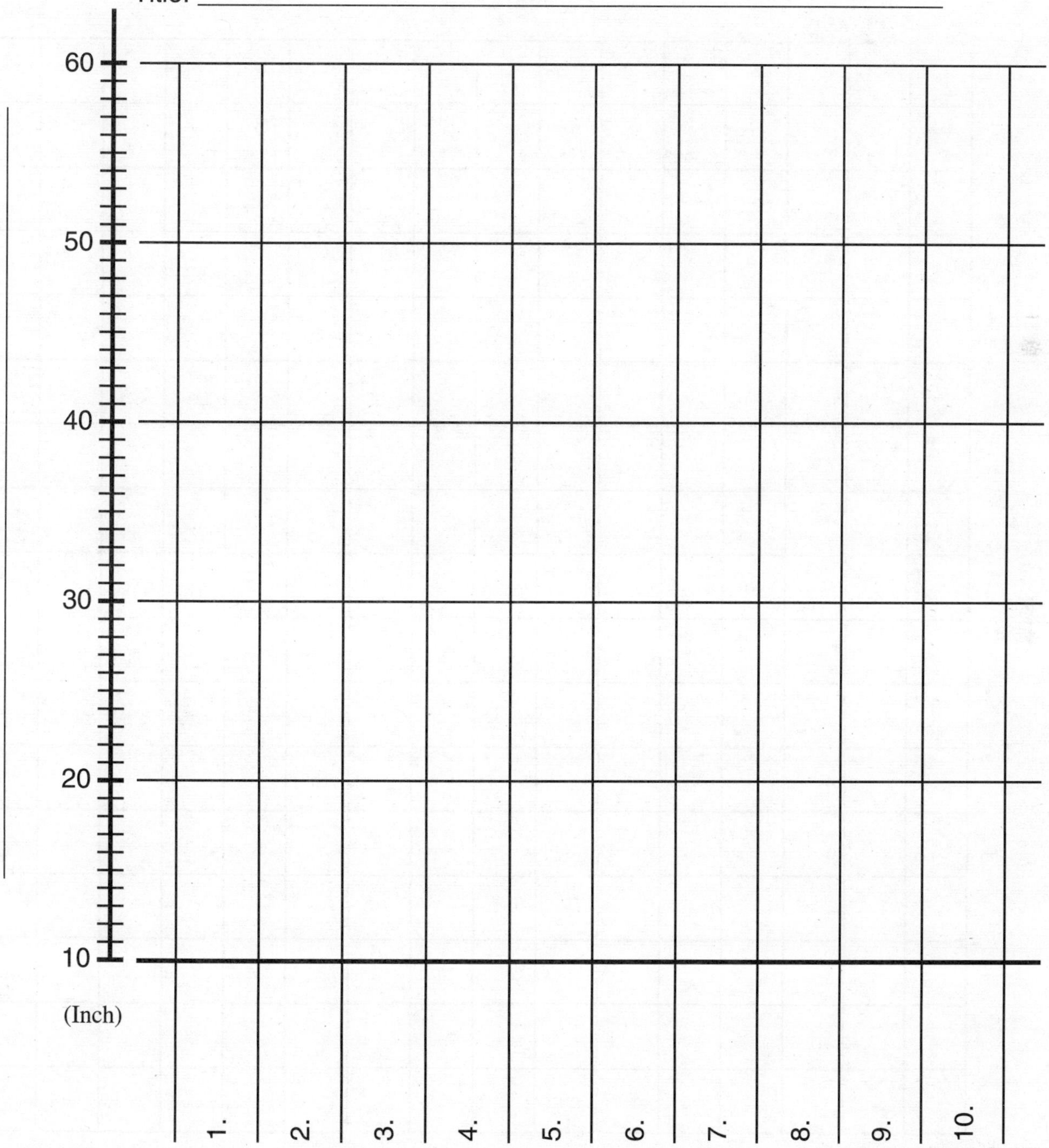

Graphing Height
身高图

Directions: Use this grid paper（网格纸） to create a bar graph of your height and your partner's height.

____

(Title)

____

We All Scream for Ice Cream

我们都要冰激凌

■ = Chocolate　■ = Strawberry　□ = Vanilla（香草）

We All Scream for Ice Cream *(cont.)*

What Is the Problem?

A teacher wants to buy her students ice cream as a classroom reward（奖励）. She only wants to buy one flavor（口味） of ice cream. She surveyed（调查） the students to find out what flavor of ice cream they like the most and how they prefer it served: in a cup or a cone（愿意吃杯装的还是吃甜筒装的）. Your job is to help the teacher decide what flavor of ice cream to buy and whether to serve it in a cup or a cone.

Use page 111 and the information on this page to answer the questions. Before you begin solving each problem, be sure to locate the key information you will need. Think about how you will draw your diagram and what you will include on it.

Remember: If you have trouble planning ways to solve a problem, reread it and rethink your strategy. When you have a solution, think about whether your answer makes sense.

Problem-Solving Strategy: Drawing a Diagram

Problem A

Using the pictures on page 111, draw a diagram to figure out how many children like chocolate ice cream, how many children like vanilla ice cream, and how many children like strawberry ice cream.

Problem B

What flavor is liked the most? What flavor is liked the least? Do most children prefer their ice cream in a cup or a cone?

Problem C

How many children like chocolate ice cream served on a cone? How many children like strawberry ice cream in a bowl?

Extension

Survey your class to find out the favorite ice cream flavor and way to serve it. Organize the information in a diagram. Discuss（讨论） the results（结果） as a class.

Standardized Test Preparation 27

标准考前测试 27

1 624 + 53 = ______

Ⓐ 677

Ⓑ 676

Ⓒ 629

Ⓓ 628

2 Which of these containers（容器）would hold less than 1 cup of liquid?

Ⓕ car gas tank（汽车油箱）

Ⓖ soda can（苏打水罐）

Ⓗ medicine cup（医用水杯）

Ⓙ drinking glass（饮水杯）

3 Which produces the same answer as 8 x 7?

Ⓐ 7 + 8

Ⓑ 8 ÷ 7

Ⓒ 8 – 7

Ⓓ 7 x 8

4 Which of these shapes has more than 7 corners?

Ⓕ pentagon（五边形）

Ⓖ hexagon（六边形）

Ⓗ octagon（八边形）

Ⓙ square（正方形）

5 Which number belongs outside the circle?

Numbers 41–50

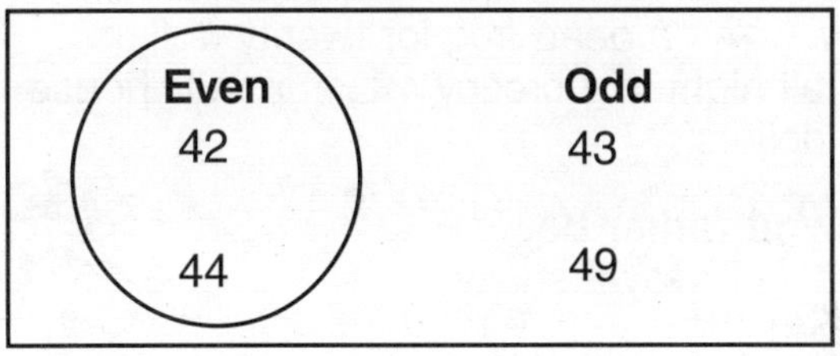

Ⓐ 46　　Ⓒ 53

Ⓑ 41　　Ⓓ 50

6 Pick one question from this test. Explain how and why you chose your answer.

____

____

____

____

____

____

Would You Rather . . .

你喜欢……

Would you rather. . .
an elephant drank（喝） your bath water（洗澡水）
an eagle stole（偷） your dinner（晚餐）
a pig tried on your clothes（试穿你的衣服）
or a hippo（河马） slept in your bed

Would you rather be. . .
covered in jam（被果酱包裹）
soaked in water（浸在水里）
or pulled through the mud（在泥地里拖）
by a dog

Would you rather have. . .
dinner in a castle（城堡）
breakfast in a balloon（气球）
or tea on the river

Would you rather be made to eat. . .
spider stew（炖蜘蛛）
slug dumplings（鼻涕虫团子）
mashed worms（捣碎的虫子）
or drink snail squash（压扁的蜗牛）

Would you rather. . .
jump in the nettles（荨麻） for five dollars
swallow（吞） a dead frog for twenty dollars
or stay all night in a creepy（毛骨悚然） house
for fifty dollars

Would you rather be. . .
crushed（挤） by a snake
swallowed（被……吞下） by a fish
eaten by a crocodile（鳄鱼）
or sat on by a rhinoceros（犀牛）

Would you rather be chased（追逐）**by. . .**
a crab（螃蟹）
a bull（公牛）
a lion（狮子）
or wolves（狼群）

Would you rather be lost. . .
in the fog（雾中）
at sea
in a desert（沙漠）
in a forest（森林）
or in a crowd（人群中）

Would you rather live with. . .
a gerbil（沙鼠） in a cage（笼子）
a fish in a bowl（缸）
a parrot（鹦鹉） on a perch（栖木）
a rabbit in a hutch（笼）
chickens in a coop（鸡笼）
or a dog in a kennel（狗屋）

from *Would You Rather...*
by John Burningham

What Would You Rather Do?

你喜欢做什么？

Directions: Write one choice above each box from the list on the board. Fill in your hypothesis（假设）about which choice you think will be most popular. Then ask students which choice they would rather do. Tally（记录）the results in the boxes. Finally, write if your hypothesis is correct or incorrect, and explain what you found out.

My hypothesis is … ____

____.

My hypothesis was *correct/incorrect*: ______________________

I found out that ____

____.

Would You Rather . . . Bar Graph

你喜欢……条形图

Directions: Complete the bar graph by writing the choices from "What Would You Rather Do?" below each bar. Then, color the bars to show how many students chose each option（选项）.

Title: ______________________________

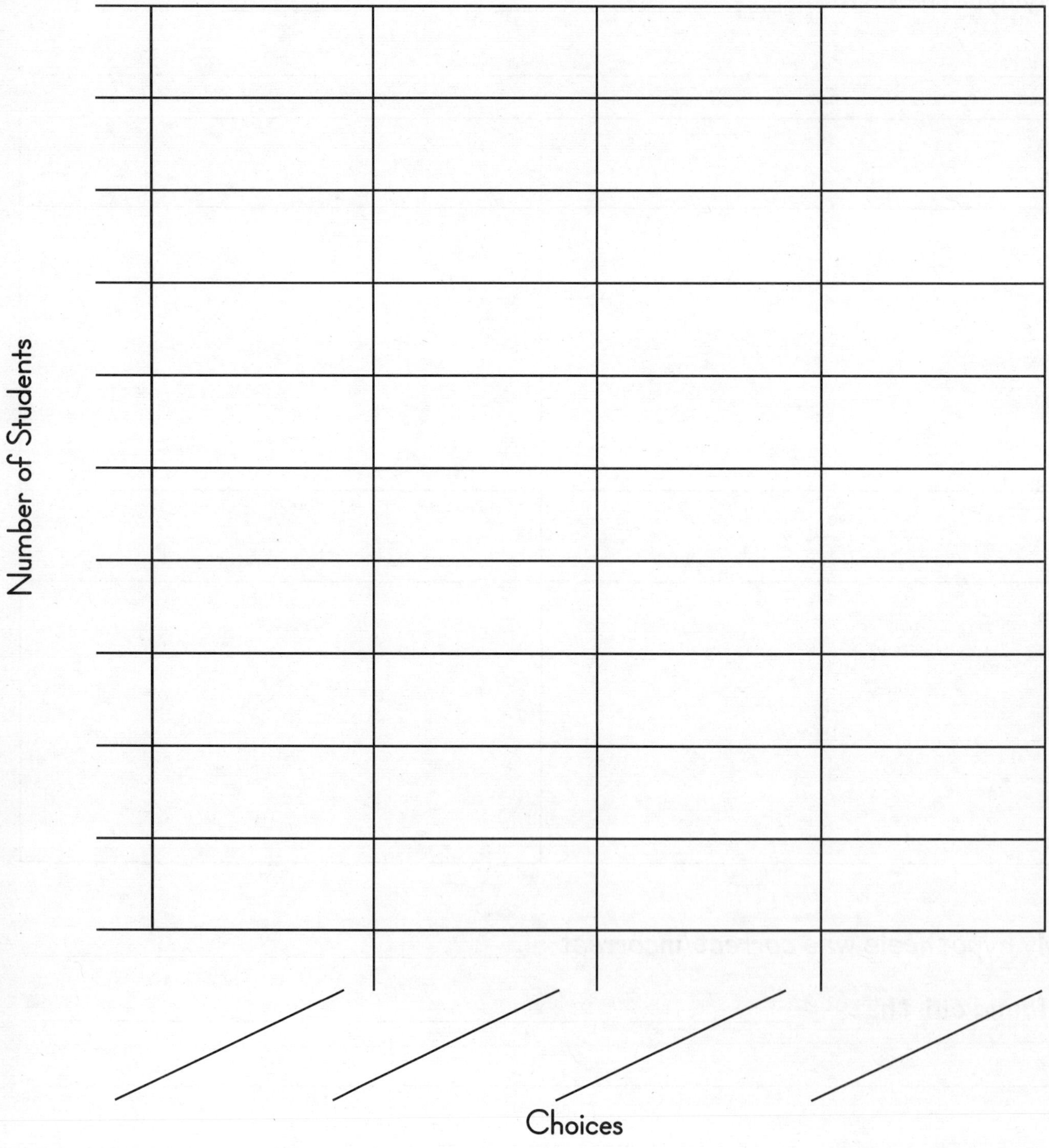

Drawing a Diagram Problems

画图解题 练习

Directions: Solve each problem and show your work.

Problem A

Four friends live in the Tall Tower apartment（公寓）building. The building is four stories（四层）high and has four apartments on each floor（楼层）. Draw a diagram to show where each of the friends live.

1. Jim lives on the 4th floor in the apartment second from the left.
2. Robin lives down three floors from Jim and over one apartment to the left.
3. Billy lives two floors above Robin and over three apartments to the right.
4. Cheryl lives two floors below Jim and in the first apartment to the left of Billy's.

Problem B

Shade each pizza（披萨饼）below to show the following fractions. Draw a line to connect the pizzas that are equal.

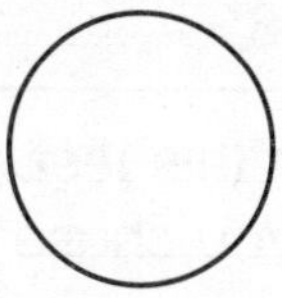

1. $\frac{2}{2}$

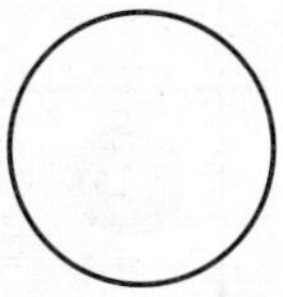

2. $\frac{1}{2}$

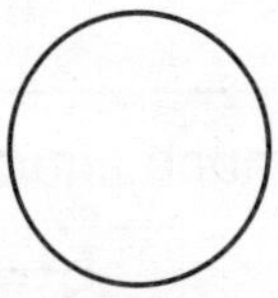

3. $\frac{3}{6}$

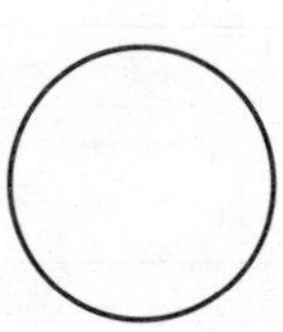

4. $\frac{2}{8}$

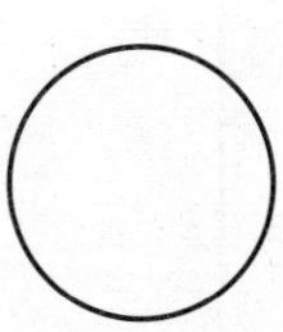

5. $\frac{4}{10}$

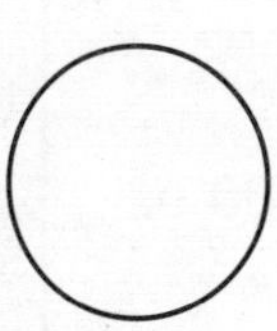

6. $\frac{1}{4}$

Standardized Test Preparation 28
标准考前测试 28

1 How many students chose orange as their favorite color?

3rd Graders' Favorite Colors

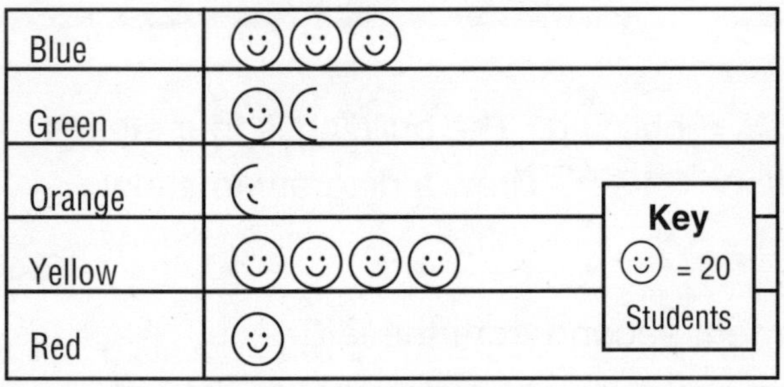

Ⓐ 1 Ⓒ 10
Ⓑ 5 Ⓓ 20

2 How many lines of symmetry can be drawn on this rectangle?

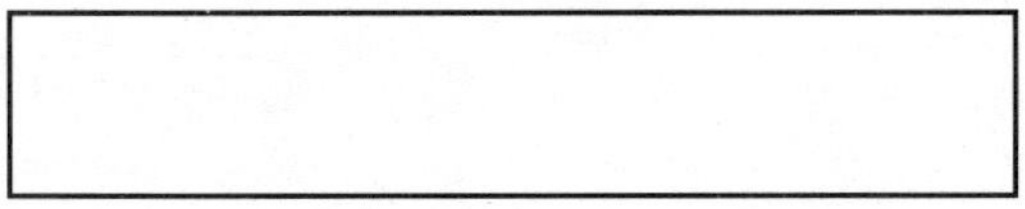

Ⓕ 1 Ⓗ 2
Ⓖ 3 Ⓙ 4

3 How much liquid is in this jug?

Ⓐ 10 oz.
Ⓑ 8 oz.
Ⓒ 5 oz.
Ⓓ 100 oz.

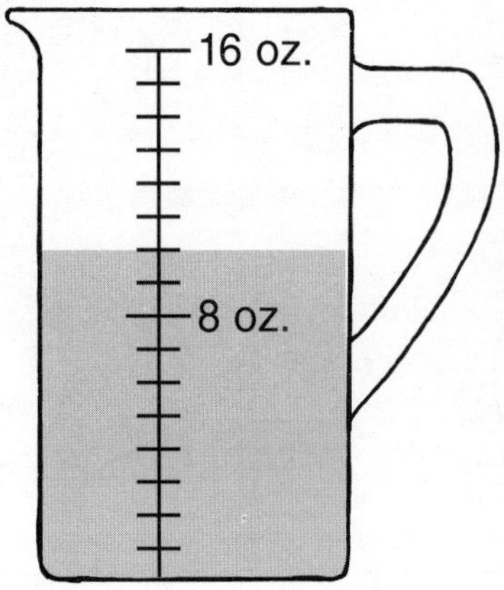

4 Which of the following statements is likely?

Ⓕ You will eat 100 candy bars（棒棒糖）today.

Ⓖ You will find a million dollars in the street.

Ⓗ You will play with a friend some time today.

Ⓙ You will fly to Mars today.

5 How many more books did Grade 2 read than Grade 4 during the first week of school?

Books Read During the First Week of School

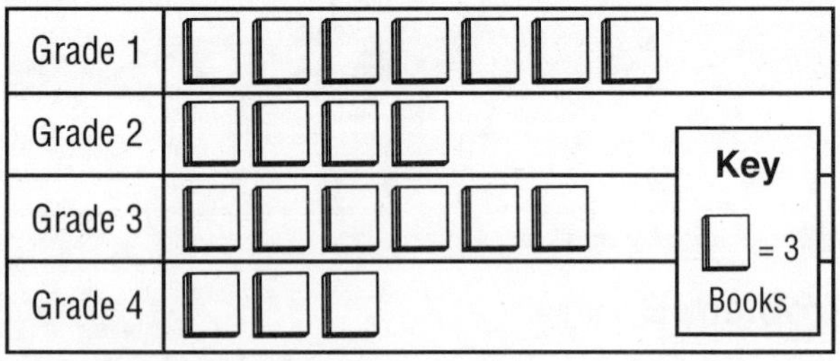

Ⓐ 1 Ⓒ 5
Ⓑ 3 Ⓓ 6

6 Pick one question from this test. Explain how and why you chose your answer.

____
____
____
____
____
____

Collecting Box Coupons

收集优惠券

Directions: Students at Newhall School have been collecting box coupons from breakfast cereals（谷类食品） to get computer equipment for their school. Students in each class made a graph to show the number of box coupons they had collected.

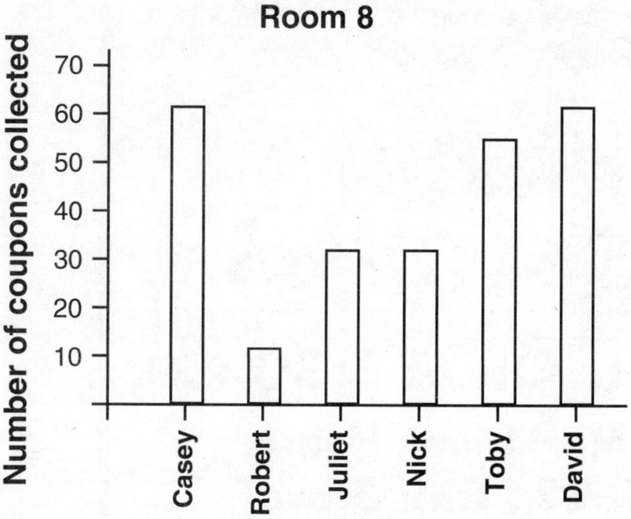

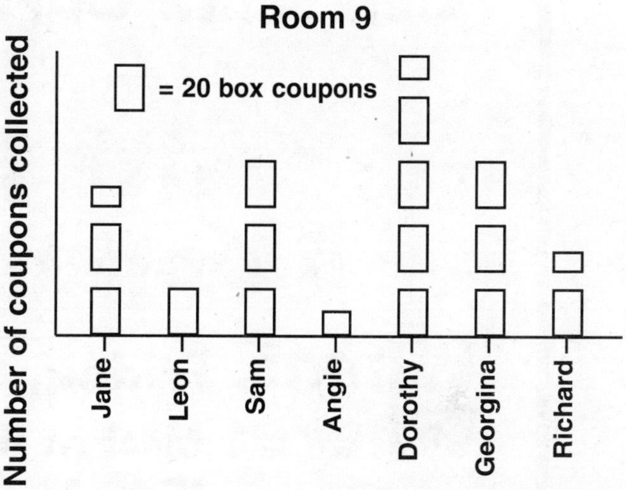

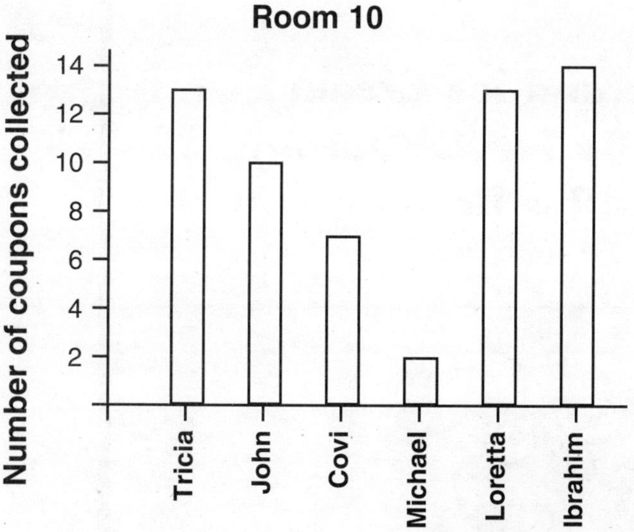

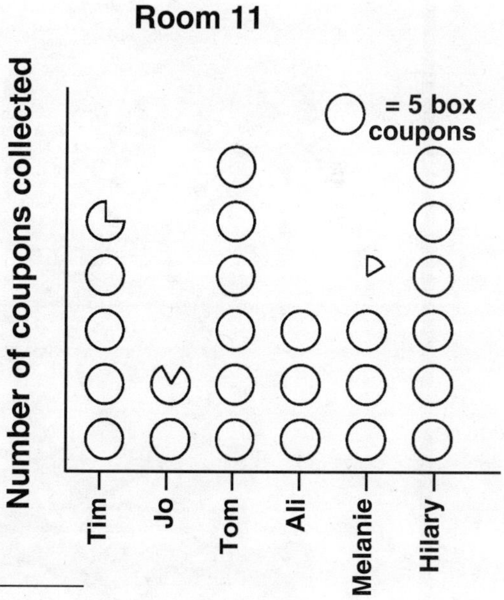

1. How many students collected box coupons? ____________
2. Which student collected the most box coupons? ____________
3. Approximately how many box coupons were collected altogether? ____________
4. How many more box coupons did Casey collect than Sam? ____________
5. How many more box coupons did Ibrahim collect than Covi? ____________
6. If 15 box coupons are needed for one CD, approximately（大致地） how many CDs have the students earned? ____________

Celebrating a Birthday

庆祝生日

Celebrate Curt's 9th Birthday

with Monster Bumper Bowling!

When: Friday, Oct. 4th at 3:30 P.M.

Where: Meet at the Markham's House, 1542 Main Street, Oak Grove

Cake and Ice Cream Will Be Served During and After Bowling!

Celebrating a Birthday *(cont.)*

What Is the Problem?

It is your birthday. Your job is to cut and serve（供应，招待）the cake. There are four people at your party. You cut the cake into eight pieces.

Use the information on this page to answer the questions. Before you begin solving each problem, be sure to locate the key information you will need. Think about how you will draw your diagram and what you will include on it.

Remember: There are hints to help you get started. If you have trouble planning ways to solve a problem, reread it and rethink your strategy. When you have a solution, think about whether your answer makes sense.

Problem-Solving Strategy: Drawing a Diagram

Problem A

Each person eats one piece of cake. How much of the cake will be eaten?

Hint: Draw a rectangle（长方形）to represent（代表）the cake. Draw lines to represent the pieces.

Problem B

Each person eats $\frac{1}{4}$ of the cake. How many pieces of cake will each person get to eat?

Hint: Draw a rectangle to represent the cake. Draw lines to represent the pieces. Shade each piece eaten with a different color.

Problem C

You want to save $\frac{1}{2}$ of the cake to eat tomorrow. What fraction of the cake will each person get to eat today?

Hint: Draw a rectangle to represent the cake. Shade the portion（部分）of the cake to be saved and color each piece eaten.

Extension

Figure out how many cakes would have to be made so that each person in your class could have a piece of cake that is $\frac{1}{4}$ of a whole（整个的）cake.

Standardized Test Preparation 29

标准考前测试 29

1 How many bananas did monkeys eat on Monday?

Number of Bananas Eaten by Monkeys

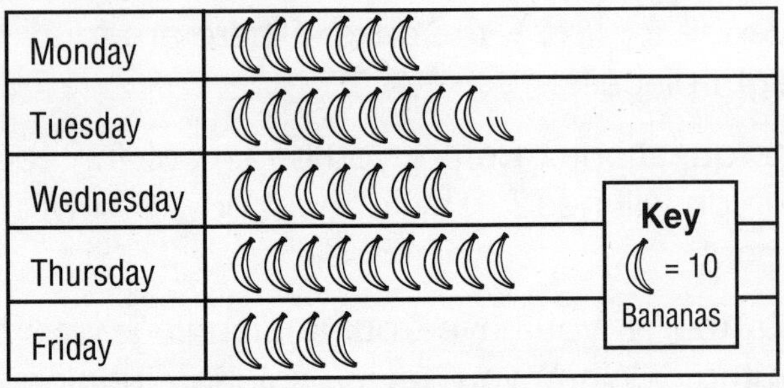

Ⓐ 600 Ⓑ 60 Ⓒ 70 Ⓓ 90

2 It takes 2 hours for the bus to travel from the beginning to the end of its route（路线，行程）. The bus began its route at 8:10. What time will the bus end its route?

Ⓕ 6:10

Ⓖ 8:20

Ⓗ 8:12

Ⓙ 10:10

3 How many lines of symmetry can be drawn on a square?

Ⓐ 1

Ⓑ 2

Ⓒ 3

Ⓓ 4

4 Which of these shapes has 4 sides and 4 corners?

Ⓕ hexagon（六边形）

Ⓖ circle（圆形）

Ⓗ quadrilateral（四边形）

Ⓙ octagon（八边形）

5 How many more students take the bus to school than ride bikes?

How Students in Our Class Get to School

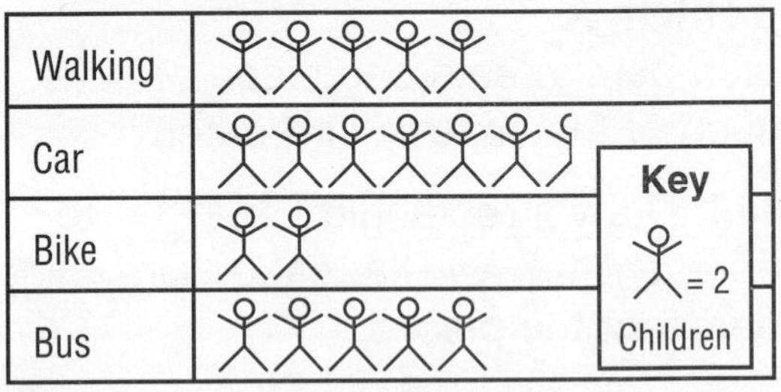

Ⓐ 3 Ⓑ 4 Ⓒ 6 Ⓓ 10

6 Pick one question from this test. Explain how and why you chose your answer.

____

____

____

____

____

____

Lesson 30

Will It or Won't It?
会不会发生？

Directions: Use the following probability scale（概率尺度）to answer the questions below.

impossible — unlikely — likely — certain

1. What does a probability scale show?

2. Place the letter of each of the following events on the probability scale above.

   a. You will be 10 on your next birthday.

   b. You will be a mother when you grow up.

   c. One member of your class will be a professional athlete（职业运动员）.

   d. Your teacher will run the Boston Marathon（波士顿马拉松赛）.

   e. One day you will run the Boston Marathon.

   f. You can say the alphabet（字母表）forwards.

   g. You can say the alphabet backwards（反向的）.

   h. You will see a zebra on the playground at lunchtime.

Challenge

Directions: Write a series of chance statements. Then, exchange papers with a classmate. Draw a probability scale, and decide where to place your classmate's statements on the scale. When finished, compare your results.

Drawing a Diagram Group Problems

画图解题 分组练习

Directions: Solve Problem A and show your work. Using the strategy you've studied this week, write and solve your own problems for Problem B and Problem C.

Problem A

Draw a shape with five sides and a perimeter of 10 inches（英寸, 1英寸=2.54厘米）.

Problem B: ______________________

____

____

____

____

____

Show work:

Answer: ______________________

Problem C: ______________________

____

____

____

____

____

Show work:

Answer: ______________________

Standardized Test Preparation 30

标准考前测试 30

1 Danica is looking for the drawer（抽屉） with more than 15 CDs but fewer than 40 CDs. Which drawer is it?

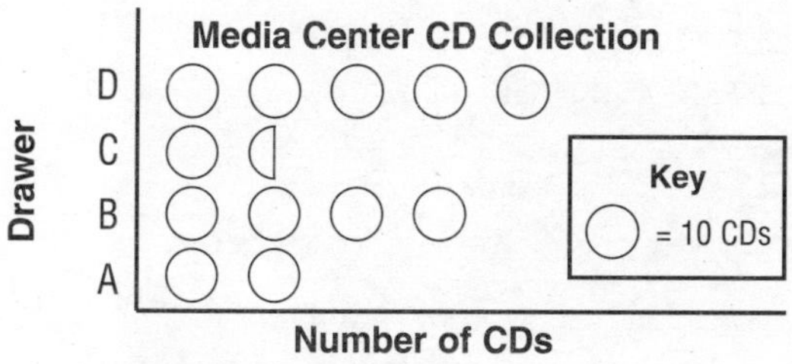

Ⓐ Drawer A　　Ⓒ Drawer C

Ⓑ Drawer B　　Ⓓ Drawer D

2 How many lines of symmetry can be drawn on this parallelogram（平行四边形）?

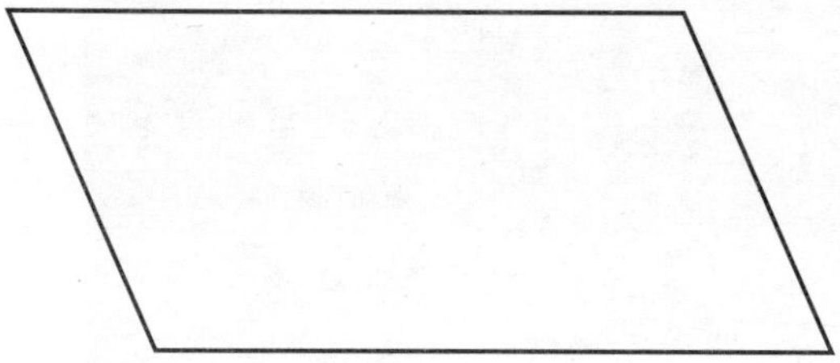

Ⓕ 0　　Ⓗ 1

Ⓖ 2　　Ⓙ 4

3 How many ounces are in this jug?

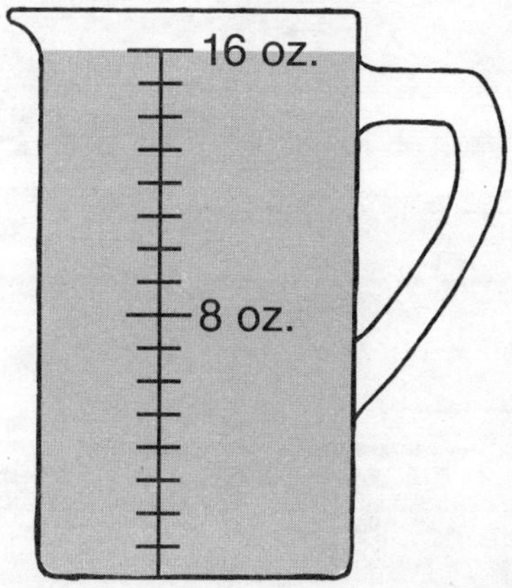

Ⓐ 1 oz.

Ⓑ 2 oz.

Ⓒ 10 oz.

Ⓓ 16 oz.

4 How many bananas did the monkeys eat on Wednesday?

Number of Bananas Eaten by Monkeys

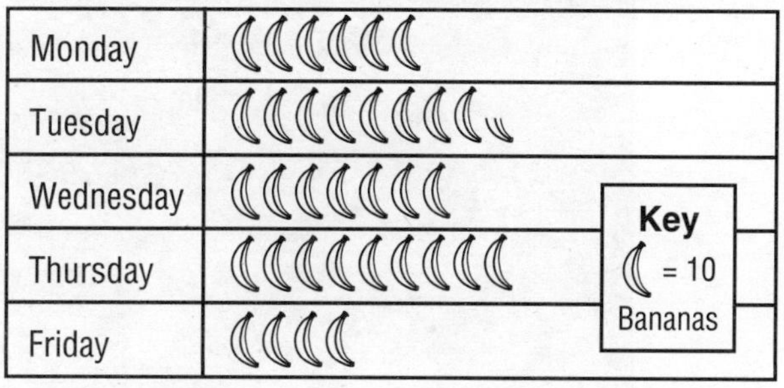

Ⓕ 35　　Ⓗ 70

Ⓖ 700　　Ⓙ 7

5 How would Ralph figure out how many pairs of pants（……条裤子） he has?

Ralph's Clothes

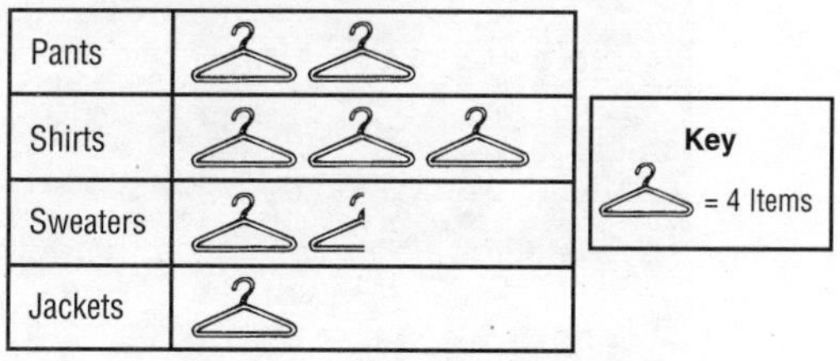

Ⓐ Add 4 to the number of .

Ⓑ Divide the number of by 4.

Ⓒ Subtract 4 from the number of .

Ⓓ Multiply 4 by the number of .

6 Pick one question from this test. Explain how and why you chose your answer.

____

____

____

____

____

____

Game Directions 游戏指南

Spin It! Directions

大转盘！游戏指南

Spin It! Directions *(cont.)*

What You Need

- *Solar System Race*（太阳系竞赛） game board
- spinner（转盘） (Divide the spinner into 8 parts and write the numbers 1–8 on it.)
- *Spin It!* cards（卡片见答案页）
- *Game Counters*（游戏计数器） or something else to use as game markers（标记物）
- pencils and paper

Object of the Game

- **Regular level play**（普通级玩法）**:** Be the first player to land（登陆） your ship on the moon.
- **Challenge level play**（挑战级玩法）**:** Be the first player to travel to the moon and back home to Earth.

Setting Up the Game

- Place the game board in the middle of all the players.
- Shuffle（洗牌） the 18 cards and place them facedown in a pile（正面朝下摞成一摞） within easy reach of all the players.
- Each player picks a rocket ship（火箭船） color. Place a game marker on the starting spot for that color.
- Decide whether to play the regular level or challenge level.

How to Play the Game

- The oldest player goes first. Then, play passes to the left.
- For each turn, draw the top card from the deck.
- Follow the directions on the card to set up a mathematics problem.
- Everyone solves the problem.
- Compare your answers. Decide who solved the problem correctly.
- Everyone who has solved the problem correctly gets to move forward one space（前进一步） on his or her space route. (**Challenge level:** If your answer is not correct, you have move back one space on the board.)
- Place the card in a discard（被打出的牌） pile. Once you run out of cards, shuffle the discard pile and use the cards again.

How to Win the Game

- To win, it has to be your turn. You cannot move past the last rocket on someone else's turn.
- **Regular level:** Once you land on the moon, you have to solve the problem below.
- **Challenge level:** Once you land on the moon, turn around and return to Earth following the steps below. Once you land on Earth, you have to solve the problem below.
  Spin three times. Create two different three-digit numbers. Find the sum of the numbers. Have your friends check your answer. If you get it correct, you win!

Match It! Directions

对对碰！游戏指南

Match It! Directions *(cont.)*

What You Need

- *Pop, Pop, Popping*（米花砰砰爆出来） game board
- spinner (Divide the spinner into 4 parts and write the numbers 1–4 on it.)
- *Match It!* cards
- *Game Counters* or something else to use as game markers
- pencils and paper

Object of the Game

- Use your memory（记忆） to match（配对） mathematical problems and their answers. Be the first player
  to move up through the kernels（谷粒） of popcorn to the piece marked **End**.

Setting Up the Game

- Place the game board in the middle of all the players.
- Shuffle the 36 cards and place them facedown. Arrange them in 6 rows（排, 行） with 6 cards in each row.
- Each player places a game marker on the kernel marked **Start**.

How to Play the Game

- The youngest player goes first. Then, play passes to the left.
- For each turn, flip over（快速翻转） two cards. Make sure that everyone can see the cards when you
  have flipped them.
- If you flip a mathematics problem, solve it. If you flip an answer, remember where it is. You are trying to match problems with their answers.
- If you don't have a match, your turn is over.
- If you make a pair, spin the spinner and move forward that many spaces. Follow any directions on the game board (move back or move ahead). Your turn is over after you move.
- Keep any pairs you find.

How to Win the Game

- The first player to land on the space labeled **End** wins.
- Second place goes to whoever has the most matches. (If the winner had the most matches, there is no second place!)

Find It! Directions

找找看！游戏指南

Find It! Directions *(cont.)*

What You Need

- *Climbing Denali*（攀登德纳里峰） game board
- spinner (Divide the spinner into 4 parts and write the numbers 1–4 on it.)
- *Find It!* cards
- *Game Counters* or something else to use as game markers
- pencils and paper

Object of the Game

- Be the first player to climb up the mountain to the **Peak**（顶峰）.

Setting Up the Game

- Place the game board in the middle of all the players.
- Shuffle the 18 question cards and place them facedown in a pile within easy reach of all the players.
- Shuffle the 18 answer cards and deal（发牌） three cards to each player.
- Each player places a game marker on the space below the mountain trail（小径）.

How to Play the Game

- The player whose birthday is closest to today goes first. Then, play passes to the left.
- For each turn, flip over a question card.
- Everyone has to solve the problem.
- Whoever has the correct answer card spins the spinner and moves forward that many spaces. Follow any directions on the game board (move back or move ahead).
- Continue playing until someone gets to the peak!

How to Win the Game

- The first player to land on the space labeled **Peak** wins.

Solve It! Directions

答题王！游戏指南

Solve It! Directions *(cont.)*

What You Need

- *Shopping Spree*（购物狂欢） game board
- spinner (Divide the spinner into 4 parts and write the numbers 1–4 on it.)
- *Solve It!* cards
- *Game Counters* or something else to use as game markers
- pencils and paper

Object of the Game

- **Regular level play:** Be the first player to circle the stores（绕店一周） and exit the mall（购物中心）.
- **Challenge level play:** Be the first player to spend $15.00 on a shopping spree.

Setting Up the Game

- Place the game board in the middle of all the players.
- Shuffle the 18 cards and place them facedown in a pile within easy reach of all the players.
- Each player places a game marker on the **Start** space.
- Decide whether to play the regular level or challenge level.

How to Play the Game

- The oldest player goes first. Then, play passes to the left.
- For each turn, draw the top card from the deck.
- Everyone solves the problem.
- Compare your answers. Decide who solved the problem correctly.
- Everyone who has solved the problem correctly gets to spin the spinner and move. Follow any directions on the game board (move back or move ahead).
- **Challenge level:** Every square you land on is a purchase（购买） you make. Create a list to record what you buy and how much it costs. You have to continue circling the mall until you spend $15.00.
- Place the card in a discard pile. Once you run out of cards, shuffle the discard pile and use the cards again.

How to Win the Game

- **Regular level:** The first player to circle the whole mall and move through the exit wins.
- **Challenge level:** The first player to "buy" (or land on) $15.00 worth of items at the mall wins.

Mathematics Chart
数学用表

Three-Dimensional Shapes（立体图形）

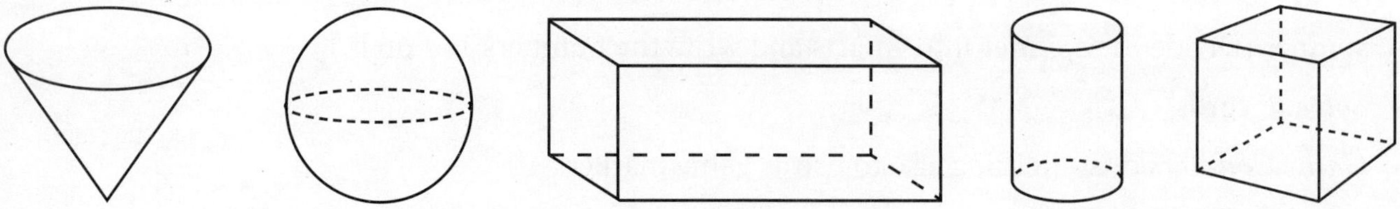

cone sphere rectangular prism cylinder cube

Two-Dimensional Shapes（平面图形）

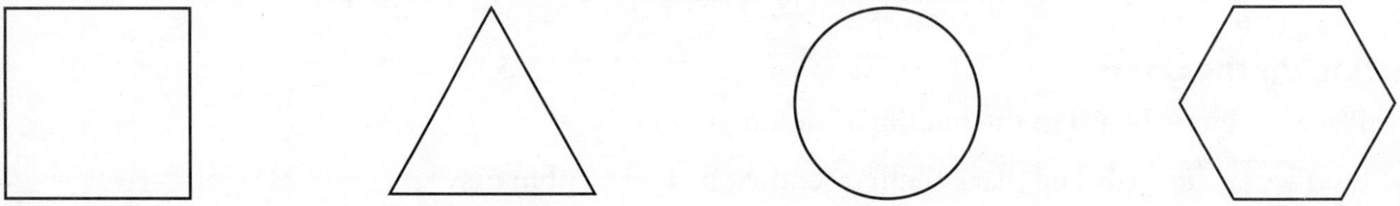

square triangle circle hexagon

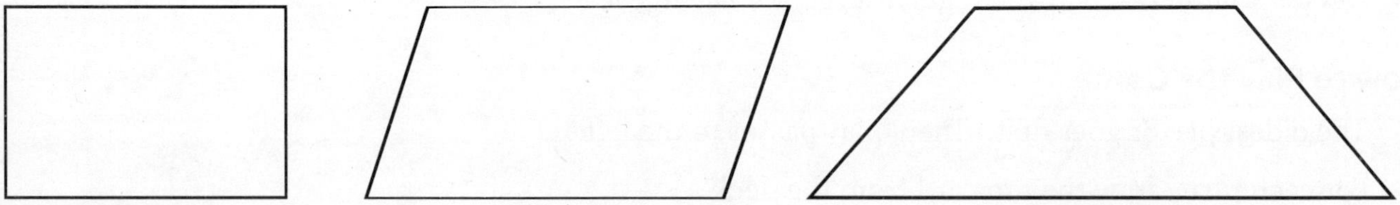

rectangle rhombus trapezoid

Length（长度）	**Measurement**（丈量）
1 foot = 12 inches	perimeter = length + length + width + width
1 yard = 3 feet	or
1 meter = 100 centimeters	perimeter = side + side + side + side
1 centimeter = 10 millimeters	area = length x width
1 foot = 0.305 meter	

Mathematics Chart *(cont.)*

Time（时间）	**Money**（钱）
1 year = 365 days	1 dollar = 100 pennies
1 year = 12 months	1 dollar = 4 quarters
1 year = 52 weeks	1 dollar = 10 dimes
1 week = 7 days	1 dollar = 20 nickels
1 day = 24 hours	1 quarter = 25¢
1 hour = 60 minutes	1 dime = 10¢
1 minutes = 60 seconds	1 nickel = 5¢
	1 penny = 1¢

Capacity (Volume)［容量（体积）］	**Weight**（重量）
1 gallon = 4 quarts	1 pound = 16 ounces
1 quart = 2 pints	1 kilogram = 1,000 grams
1 pint = 2 cups	1 gram = 1,000 milligrams
1 cup = 8 ounces	1 pound = 0.454 kilogram
1 liter = 1,000 milliliters	

Fractions（分数）

$\frac{1}{2}$ ← numerator (1), ← denominator (2)

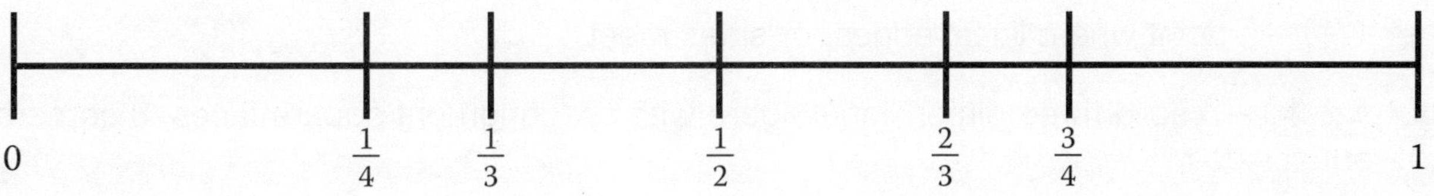

Glossary
词汇表

analog（模拟式）—a clock that has rotating hour and minute hands to show the time

approximate（近似）—to estimate a number, amount, or total; to come near

array（排列）—the arrangement of items or objects in equal rows

asymmetrical（非对称的）—not symmetrical; does not have sides/proportions exactly the same

axis（轴）—a real or imaginary reference line

axis of symmetry（对称轴）—symmetry in relation to an axis; divides a symmetrical object in half

calculate（计算）—to determine using mathematical processes

capacity（容量）—see *volume*

centimeter（厘米）—a measurement in the metric system; 0.39 of an inch

column（列）—a vertical arrangement of items written on a page

Commutative Property［交换性（加法的交换性质，交换加数，其和不变）］—(Commutative Property of Addition)—the sum remains the same when the order of the addends is changed; $a + b = b + a$, where a and b are any real numbers—Example: $5 + 2 = 2 + 5$

compare（比较）—to examine items to discover similarities or differences

cone（圆锥体）—a solid, three-dimensional shape with an elliptical or circular base and a curved surface that tapers to a point (the vertex)

congruent（全等的，全等号为"≅"）—having the same size and shape; the symbol for congruence is ≅

corner（角）—a point where lines, edges, or sides meet

cube（立方体）—a solid three-dimensional figure with six congruent square faces, 8 corners, and 12 equal sides

cup（杯，半品脱，8盎司）—a half pint; eight fluid ounces

curved（弯曲的）—something that is not straight

cylinder（圆柱体）—a solid, three-dimensional shape with a curved surface and two congruent elliptical or circular bases

data（数据）—factual information used as a basis for reasoning, discussion, and calculation

decreasing（递减的，减少的）—becoming progressively less in size, amount, or number

Glossary *(cont.)*

degree（度，温度的常用单位）—a customary unit for measuring temperature

denominator（分母）—the quantity below the line in a fraction

digit（数字）—any of the ten numerals 0 through 9

digital（数字时钟）—a clock that uses only numerals to show the time

dimension（维）—refers to length, breadth, and height

distance（距离）—the length between two points, lines, or objects

division（除法）—the operation of creating equal groups

double（使加倍）—to make twice as great

edge（边）—line segment where two faces of a solid figure meet

equal（等于）—of the same value as another

equation（等式，方程式）—a mathematical statement where the left side of the equal sign has the same value as the right

equivalent（等值的）—equal in value

estimate（估算）—to make a rough calculation

even number（偶数）—a number that is divisible by two

expanded notation（计值法）—a way to write numbers as the sum of the values of each digit

foot（英尺，长度单位，1英尺=12英寸）—a unit of length measurement; 12 inches

fraction（分数）—a number that identifies part of a whole or part of a group

fractional parts（分数部分）—parts of a fraction

greater than（大于，用“>”表示）—more than; shows relationship between numbers; >

grid（坐标方格）—a network of horizontal and vertical lines

half（一半）—one of two equal parts that together make a whole

height（高）—the distance from the bottom to the top of something

horizontal axis（水平轴，横轴）—an axis displayed along the horizon

Glossary *(cont.)*

hypothesis (假设)—a tentative assumption made before testing its validity

inch (英寸)—a unit of length measurement; $\frac{1}{12}$ of a foot

increasing (递增的)—becoming progressively more in size, amount, or number

increment (增加)—the amount or degree that something changes

interpret (解释, 说明)—to explain or tell the meaning of

inverse operation (逆运算)—opposite, reverse operations; example—addition and subtraction

inverse relationship (反相关)—the relationship between two inverse operations

length (长度)—a measured distance along a line or figure from one end to the other

less than (小于, 少于)—not as many as; shows a relationship between numbers; $<$

maximum (最大值)—the highest or greatest quantity or value attainable

milliliters (ml) (毫升)—a metric unit for measuring capacity; equal to one-thousandth of a liter

minimum (最小值)—the lowest or smallest quantity or value attainable

multiple (倍乘)—the product of a number and any other number

multiples (倍数)—repeated groups of the same amount

number line (数字直线)—a diagram that shows numbers as points on a line

number sentence (算式)—a sentence that includes numbers, operation symbols, and a greater than or less than symbol or an equal sign

numerator (分子)—the number in a fraction that is above the line

numerical scale (数值标度)—the basis of a number system

odd number (奇数)—a whole number that is not divisible by two

ounce (盎司)—a unit of weight measurement; $\frac{1}{16}$ of a pound

part (份额, 一份)—one of the units that something can be divided into

pattern (模式, 规律)—a form or model by which elements can be arranged so that what comes next can be predicted

Glossary *(cont.)*

perimeter （周长）—the distance around a figure

picture graph （图表）—a graph using pictures or symbols to represent quantity

place value （数位值）—the value of the place of a digit in a number

prism （棱柱）—a three-dimensional figure with two identical, parallel bases

probability （概率，可能性）—the chance that a given event or outcome will occur; measured as a ratio of the total of possible outcomes

product （乘积）—the result when two numbers are multiplied

reflective symmetry （镜像对称）—a line that acts like a mirror across a picture and both sides look exactly the same

regroup （重新组合）—to exchange amounts of equal value to rename a number

remainder （余数）—the amount left over when a number cannot be divided equally

represent （代表）—to serve as a sign or symbol

right-angle triangle （直角三角形）—a triangle that has a right angle

rounding （四舍五入）—finding the nearest ten, hundred, tenth, hundredth, and so on, by following specific rules for rounding numbers

row （行）—a number of objects arranged in a horizontal line

scale （比例）—a proportion between two sets of measurements

sequence （序列）—an ordered set of numbers, shapes, or other objects, arranged according to a rule

set （集合）—a collection of objects; members of the set are called elements

shape patterns （图形规律）—repeated shape designs or sequences

side （边）—a bounding line or curve on the edge of a shape joining its vertices

simplified （简化）—made simple or less complex; in a fraction when the numerator and denominator are reduced to the smallest numbers possible

standard form （标准表格）—most common way of representing whole numbers, integers, and decimals

Glossary *(cont.)*

sum (和)—the results of adding numbers

symbol (标志, 符号)—a sign that represents something intangible

symmetry (对称)—having balanced proportions; one half is the same as the other

temperature (温度)—how hot or cold something is; measured in degrees using a thermometer; two common scales are the Celsius scale (°C) and the Fahrenheit scale (°F)

thermometer (温度计)—an instrument used to measure temperature

three-dimensional (三维的, 立体的)—having three dimensions: length, width, and height

total (总数, 合计)—the sum or whole amount; the result of adding

turns (转动)—rotations around a point

two-dimensional (二维的, 平面的)—having two dimensions: height and length

vertex (vertices) (顶点)—the point at which two lines meet

vertical (垂直的)—located at right angles to the horizon

vertical axis (垂直轴)—the axis at a right angle to the horizon

volume (体积)—the number of cubic units it takes to fill a figure

whole (整体, 全部)—having all of its parts

whole number (整数)—a counting number from zero to infinity

width (宽)—horizontal measurements taken at the right angles to the length

Appendix D: 附录D

Answer Key

参考答案

Diagnostic Test (pages 2–6)

1. B	2. J	3. C
4. J	5. A	6. G
7. A	8. J	9. B
10. J	11. B	12. G
13. C	14. J	15. D
16. H	17. A	18. G
19. C	20. F	21. A
22. J	23. C	24. J
25. D	26. F	27. C
28. F	29. D	30. J

Creating a Table 1 (page 8)

Extension Question

48 years old

Standardized Test Preparation 1 (page 9)

1. A 2. H 3. C 4. F
5. C 6. Answers will vary.

Number Lines (page 10)

Answers will vary. Check students' work.

Pet Contest (pages 11–12)

Problem A

first place: horse
second place: dog
third place: cat

Problem B

$60; horse

Problem C

39 mph faster

Extension

The cheetah would have come in first place. First place would receive $70.

Standardized Test Preparation 2 (page 13)

1. D 2. J 3. B 4. G
5. C 6. Answers will vary.

Make the Number (page 14)

1.

9	3	14
5	7	2
12	15	8

2.

9	15	4
13	11	16
6	3	10

3.

91	85	96
87	89	84
94	97	90

4.

71	65	76
67	69	64
74	77	70

5.

21	15	26
17	19	14
24	27	20

6.

12	12	12
12	12	12
12	12	12

Addition Problems (page 15)

Part A

1. 42	2. 72
3. 63	4. 97
5. 49	6. 124
7. 115	8. 117

Part B

9. 658	10. 174
11. 496	12. 232
13. 461	14. 287
15. 244	16. 927

Part C

17. 1,234	18. 618
19. 949	20. 914
21. 620	22. 919
23. 1,047	24. 1,001

Creating a Table 1 Problems (page 16)

Problem A

10 days; $1.00 or 100¢

Day	Money Earned	Money Saved
1	10¢	5¢
2	20¢	10¢
3	30¢	15¢
4	40¢	20¢
5	50¢	25¢
6	60¢	30¢
7	70¢	35¢
8	80¢	40¢
9	90¢	45¢
10	$1.00	50¢

Problem B

18 minutes

Minutes	Apples in Basket
3	5
6	10
9	15
12	20
15	25
18	30

Standardized Test Preparation 3 (page 17)

1. C 2. F 3. B 4. G
5. C 6. Answers will vary.

Number Sacks (page 18)

Answers will vary.

Sample: 8 + 9 = 17 9 + 8 = 17
17 – 8 = 9 17 – 9 = 8

Answer Key *(cont.)*

Subtraction Problems (page 19)

Part A

1. 19
2. 27
3. 21
4. 64
5. 43
6. 19
7. 37
8. 64

Part B

9. 128
10. 364
11. 318
12. 37
13. 137
14. 206
15. 883
16. 669

Part C

17. 165
18. 315
19. 466
20. 68
21. 43
22. 377
23. 241
24. 563

Buying School Supplies (pages 20–21)

Problem A

$0.63 or 63¢

Problem B

$1.40

Problem C

1 eraser

Extension

$0.50 or 50¢

Standardized Test Preparation 4 (page 22)

1. C 2. H 3. B 4. G

5. D 6. Answers will vary.

Rounding (page 24)

Part A

1. 100
2. 100
3. 600
4. 500
5. 600
6. 500
7. 800
8. 0
9. 200
10. 1,000

Part B

11. 50 + 40 = 90
12. 120 – 50 = 70
13. 60 – 20 = 40
14. 360 + 410 = 770
15. 60 + 90 = 150
16. 450 + 570 = 1020
17. 900 – 200 = 700
18. 230 + 180 = 410

Creating a Table 1 Group Problems (page 25)

Problem A

family = $2.10; Mom = $1.05 or 105¢; Dad = 70¢; Junior = 35¢

Day	Dad	Mom	Junior	Total
Sunday	10¢	15¢	5¢	30¢
Monday	20¢	30¢	10¢	60¢
Tuesday	30¢	45¢	15¢	90¢
Wednesday	40¢	60¢	20¢	$1.20
Thursday	50¢	75¢	25¢	$1.50
Friday	60¢	90¢	30¢	$1.80
Saturday	70¢	$1.05	35¢	$2.10
Total	70¢	$1.05	35¢	$2.10

Standardized Test Preparation 5 (page 26)

1. C 2. F 3. C 4. F

5. B 6. Answers will vary.

Don't Count on It! 1 (pages 27–28)

1. 6 x 5 = 30
2. 6 x 11 = 66
3. 4 x 8 = 32
4. 8 x 9 = 72
5. 7 x 10 = 70
6. 5 x 9 = 45

Don't Count on It! 2 (page 29)

Check students' arrays

1. 12
2. 42
3. 16
4. 72
5. 35
6. 18
7. 21
8. 6
9. 24

Guessing and Checking (page 30)

Extension Question

21 + 16 = 37

Standardized Test Preparation 6 (page 31)

1. B 2. H 3. C 4. F

5. B 6. Answers will vary.

Multiplication Grid (page 32)

10	10	20	30	40	50	60	70	80	90	100
9	9	18	27	36	45	54	63	72	81	90
8	8	16	24	32	40	48	56	64	72	80
7	7	14	21	28	35	42	49	56	63	70
6	6	12	18	24	30	36	42	48	54	60
5	5	10	15	20	25	30	35	40	45	50
4	4	8	12	16	20	24	28	32	36	40
3	3	6	9	12	15	18	21	24	27	30
2	2	4	6	8	10	12	14	16	18	20
1	1	2	3	4	5	6	7	8	9	10
x	**1**	**2**	**3**	**4**	**5**	**6**	**7**	**8**	**9**	**10**

On the Farm (pages 33–34)

Problem A

195 eggs

Problem B

5 cows and 1 chicken; (5 cows x 4 legs = 20 legs; 1 chicken x 2 legs = 2 legs; 20 legs + 2 legs = 22 legs)

Problem C

50 gallons (15 cows x 10 gallons per cow = 150 gallons; wife = 100 gallons packaged)

Answer Key *(cont.)*

Extension

10 cows (24 hens x 3 eggs per day = 72 eggs; 72 eggs / 12 = 6 dozen eggs; 6 x $2.50 = $15.00; $315 – $15.00 = $300.00)

Standardized Test Preparation 7 (page 35)

1. A 2. H 3. B 4. J
5. C 6. Answers will vary.

Guessing and Checking Problems (page 37)

Problem A

Answers will vary. Sample: Top circle—7; Middle circles—8 and 9; Bottom circles—6, 10, and 5

Problem B

Answers will vary. Sample: Top row—8, 10, and 6; Middle row—4, 9, and 11; Bottom row—12, 5, and 7

Standardized Test Preparation 8 (page 38)

1. C 2. H 3. B 4. G
5. C 6. Answers will vary.

Missing Numbers (page 39)

1. 6	10. 8
2. 8	11. 6
3. 4	12. 4
4. 7	13. 8
5. 9	14. 9
6. 9	15. 3
7. 9	16. 7
8. 4	17. 5
9. 6	18. 7

Make Four (page 40)

Answers will vary. Sample answers are as follows:

1. 3, 6, 18
   3 x 6 = 18   6 x 3 = 18
   18 ÷ 3 = 6   18 ÷ 6 = 3
2. 5, 2, 10
   5 x 2 = 10   2 x 5 = 10
   10 ÷ 2 = 5   10 ÷ 5 = 2
3. 7, 4, 28
   7 x 4 = 28   4 x 7 = 28
   28 ÷ 4 = 7   28 ÷ 7 = 4
4. 9, 4, 36
   9 x 4 = 36   4 x 9 = 36
   36 ÷ 4 = 9   36 ÷ 9 = 4
5. 6, 4, 24
   6 x 4 = 24   4 x 6 = 24
   24 ÷ 6 = 4   24 ÷ 4 = 6
6. 9, 2, 18
   9 x 2 = 18   2 x 9 = 18
   18 ÷ 2 = 9   18 ÷ 9 = 2

How to Play Football (pages 41–42)

Problem A

1 touchdown; it was an extra point after a touchdown

Problem B

5 field goals

Problem C

Game 1 = 2 touchdowns, 2 extra points, & 2 field goals; Game 2 = 2 touchdowns, 2 extra points, and 1 field goal

Extension

2 touchdowns; Team A: 20 pts. Team B: 18 pts.

Standardized Test Preparation 9 (page 43)

1. A 2. G 3. B 4. G
5. A 6. Answers will vary.

Divide and Check (page 44)

1. 10	2. 8	3. 9
4. 9	5. 5	6. 9
7. 7	8. 7	9. 11
10. 4	11. 7	12. 9
13. 6	14. 11	15. 9
16. 6	17. 9	18. 7
19. 12	20. 9	

Guessing and Checking Group Problems (page 45)

Problem A

Corie = 18 years old; Aidan = 9 years old; Dillon = 6 years old; Charts will vary depending on students' guesses.

Standardized Test Preparation 10 (page 46)

1. C 2. H 3. B 4. H
5. A 6. Answers will vary.

Fractions Fun (page 47)

Answers will vary.

Color the Fraction (page 48)

1. Fraction colored $\frac{3}{8}$
   Fraction not colored $\frac{5}{8}$
2. Fraction colored $\frac{3}{4}$
   Fraction not colored $\frac{1}{4}$
3. Fraction colored $\frac{7}{10}$
   Fraction not colored $\frac{3}{10}$
4. Fraction colored $\frac{2}{3}$
   Fraction not colored $\frac{1}{3}$
5. Fraction colored $\frac{3}{5}$
   Fraction not colored $\frac{2}{5}$

Answer Key *(cont.)*

6. Fraction colored $\frac{4}{8}$ or $\frac{1}{2}$
   Fraction not colored $\frac{4}{8}$ or $\frac{1}{2}$
7. Fraction colored $\frac{5}{8}$
   Fraction not colored $\frac{3}{8}$

Looking for a Pattern (page 49)

Extension Question

Day 5; Day 8

Standardized Test Preparation 11 (page 50)

1. B 2. H 3. C 4. G
5. B 6. Answers will vary.

Fractions of a Set (page 51)

1. D 2. C 3. A
4. D 5. A 6. B
7. B 8. A

Picture Fractions (page 52)

1. 3, $\frac{1}{4}$ 2. 2, $\frac{1}{6}$ 3. 12, 1
4. check to see that correct amounts are shaded in each circle; $\frac{1}{4}$ should be circled.
5. check to see that the correct amounts are shaded in each circle; $\frac{1}{3}$ should be circled.
6. 4, $\frac{1}{3}$ 7. 6, $\frac{1}{2}$
8. check to see that the correct amounts are shaded in each circle; smallest = $\frac{1}{8}$, largest = $\frac{1}{3}$
9. 8, $\frac{1}{2}$ 10. 4; $\frac{1}{4}$

Taking a Field Trip (pages 53–54)

Problem A

25 adults; multiplying by 5

Problem B

students = 100;
total people = 125

Problem C

3 buses

Extension

Yes, room is left on the bus. Yes, the fourth graders and adults would fit.; 115 souvenirs given; 15 left-over souvenirs

Standardized Test Preparation 12 (page 55)

1. B 2. F 3. D 4. H
5. B 6. Answers will vary.

Fraction Wall (page 56)

$\frac{1}{2}$	$\frac{1}{2}$								
$\frac{1}{4}$	$\frac{1}{4}$	$\frac{1}{4}$	$\frac{1}{4}$						
$\frac{1}{5}$	$\frac{1}{5}$	$\frac{1}{5}$	$\frac{1}{5}$	$\frac{1}{5}$					
$\frac{1}{8}$	$\frac{1}{8}$	$\frac{1}{8}$	$\frac{1}{8}$	$\frac{1}{8}$	$\frac{1}{8}$	$\frac{1}{8}$	$\frac{1}{8}$		
$\frac{1}{10}$	$\frac{1}{10}$	$\frac{1}{10}$	$\frac{1}{10}$	$\frac{1}{10}$	$\frac{1}{10}$	$\frac{1}{10}$	$\frac{1}{10}$	$\frac{1}{10}$	$\frac{1}{10}$

Greater or Less Than (page 57)

1. < 2. < 3. >
4. > 5. > 6. <
7. < 8. <

Comparing Fractions (page 58)

1. < 2. < 3. >
4. > 5. > 6. <
7. > 8. >

Looking for a Pattern Problems (page 59)

Problem A

A. 65, 75, 85, 95, 105, 115, 125; add 10
B. 189, 289, 389, 489, 589, 689, 789; add 100
C. 413, 418, 423, 428, 433, 438, 443; add 5

Problem B

One more section is added to the square each time.

Standardized Test Preparation 13 (page 60)

1. B 2. H 3. A 4. F
5. A 6. Answers will vary.

Equivalent Fractions (page 61)

1. 2 2. 1 3. 2
4. 4 5. 2 6. 1
7. 4 8. 1

Ordering Fractions (page 62)

1. $\frac{1}{4}, \frac{3}{4}$ 2. $\frac{1}{3}, \frac{2}{3}$
3. $\frac{1}{6}, \frac{2}{6}, \frac{3}{6}, \frac{4}{6}, \frac{5}{6}$
4. $\frac{2}{7}, \frac{5}{7}, \frac{6}{7}, \frac{7}{7}$
5. $\frac{1}{11}, \frac{3}{11}, \frac{5}{11}, \frac{11}{11}$
6. $\frac{2}{13}, \frac{3}{13}, \frac{8}{13}, \frac{13}{13}$
7. $\frac{1}{5}, \frac{2}{5}, \frac{3}{5}, \frac{5}{5}$

Pay Day (pages 63–64)

Problem A

$9.00

Problem B

$27.00

Problem C

5 months

Answer Key *(cont.)*

Extension

$36.00; $9.00

Standardized Test Preparation 14 (page 65)

1. A 2. G 3. B 4. H
5. B 6. Answers will vary.

Patterns and Sequences (page 66)

1. 0, 2, 4, 6, 8, 10; add 2
2. 50, 45, 40, 35, 30, 25; subtract 5
3. 22, 32, 42, 52, 62, 72; add 10
4. 601, 501, 401, 301, 201; subtract 100
5. 41, 36, 31, 26, 21, 16; subtract 5
6. 98, 96, 94, 92, 90, 88; subtract 2
7. 12, 17, 22, 27, 32, 37; add 5
8. triangle
9. upright heart
10. small rectangle in top half

Looking for a Pattern Group Problems (page 67)

Problem A

A. $\frac{1}{6}, \frac{1}{7}, \frac{1}{8}, \frac{1}{9}$;
the denominator increases by one.

B. $\frac{1}{16}, \frac{1}{32}, \frac{1}{64}, \frac{1}{128}$;
the denominator doubles each time.

C. $\frac{5}{10}, \frac{6}{12}, \frac{7}{14}, \frac{8}{16}$;
the numerator increases by one and the denominator is double the numerator

Standardized Test Preparation 15 (page 68)

1. A 2. H 3. C 4. H
5. D
6. Answers will vary.

Robot (page 69)

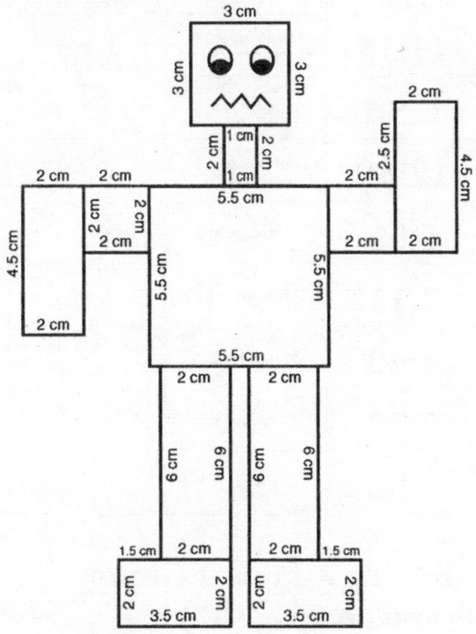

Challenge

Answers may vary.

Creating a Table 2 (page 70)

Extension Question

100 people

Standardized Test Preparation 16 (page 71)

1. C 2. H 3. D 4. H
5. A 6. Answers will vary.

Perimeters of Shapes (page 72)

1. P = 4 + 4 + 3.5 = 11.5 cm
2. P = 3.5 + 3.5 + 3.5 = 14 cm
3. P = 2.5 + 2.5 + 2.5 + 2.5 + 2.5 = 12.5 cm
4. P = 2 + 2 + 2 + 2 + 2 + 2 = 12 cm

Challenge

Answers may vary.

Car Trouble (pages 73–74)

Problem A

Estimates will vary.;
Actual length of small car = 9;
Actual height of small car = 3;
Actual length of large car = 13;
Actual height of large car = 4

Problem B

large display case = 13 cm long and 4 cm tall; small display case = 9 cm long and 3 cm tall

Problem C

4 cm longer

Extension

22 cm long

Standardized Test Preparation 17 (page 75)

1. C 2. J 3. D 4. H
5. C 6. Answers will vary.

Crazy Capacities (page 76)

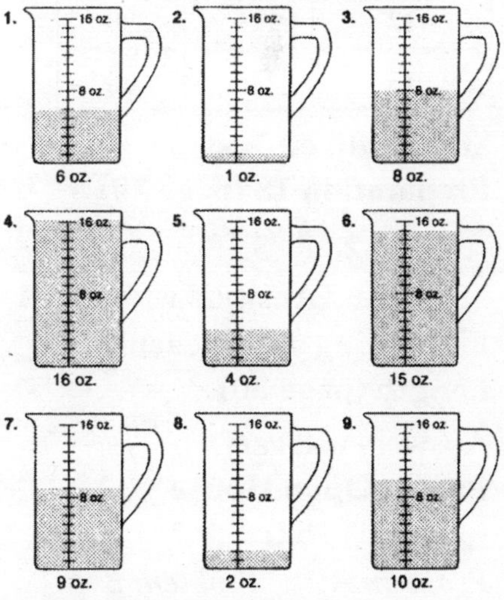

Measuring Capacities (page 77)

1. B	2. B	3. A
4. A	5. A	6. B
7. A	8. A	9. B

Answer Key *(cont.)*

Creating a Table 2 Problems (page 78)

Problem A

5 chickens and 5 sheep

Animals	Chicken Legs	Sheep Legs	Total Number of Legs
1	2	4	6
2	4	8	12
3	6	12	18
4	8	16	24
5	10	20	30

Problem B

15 strawberry pies
20 lemon pies
35 pies total

Strawberry	Lemon	Total
3	4	7
6	8	14
9	12	21
12	16	28
15	20	35

Standardized Test Preparation 18 (page 79)

1. C 2. F 3. B 4. H
5. C 6. Answers will vary.

Estimating and Measuring Length (page 80)

Answers will vary.

Parent Open House (pages 81–82)

Problem A
3 hours

Problem B
yes

Problem C
6 hours

Extension
history class starts at 1:30; science class starts at 8:30

Standardized Test Preparation 19 (page 83)

1. A 2. H 3. A 4. H
5. A 6. Answers will vary.

Digital and Analog (page 84)

Digital	Analog	Clock Face
11:15	quarter after 11; 15 minutes after 11	
6:50	10 minutes to 7	
12:45	quarter to 1; 15 minutes to 1	
5:55	5 minutes to 6	
3:11	11 minutes after 3	
4:49	11 minutes to 5	
6:34	26 minutes to 7	
4:55	5 minutes to 5	

Creating a Table 2 Group Problems (page 87)

Problem A

earns $1.00; saves 30¢; spends 70¢

Newspapers	Earns	Saves	Spends
1	10¢	3¢	7¢
2	20¢	6¢	14¢
3	30¢	9¢	21¢
4	40¢	12¢	28¢
5	50¢	15¢	35¢
6	60¢	18¢	42¢
7	70¢	21¢	49¢
8	80¢	24¢	56¢
9	90¢	27¢	63¢
10	$1.00	30¢	70¢

Standardized Test Preparation 20 (page 88)

1. B 2. H 3. B 4. F
5. C 6. Answers will vary.

Sorting Shapes (page 89)

Answers will vary. Possible answers include:

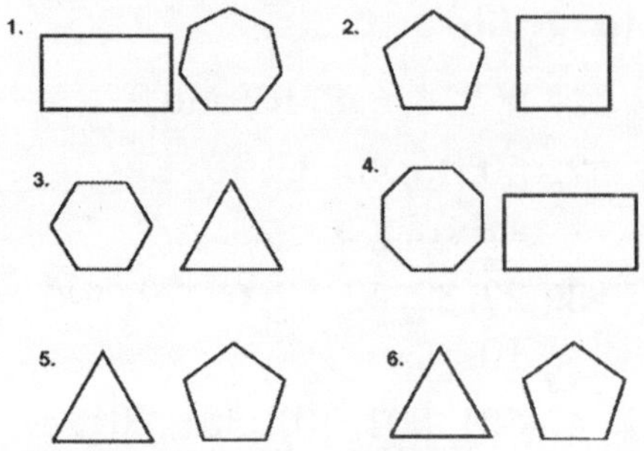

Acting It Out or Using Concrete Materials (page 90)

Answers are included on the transparency.

Standardized Test Preparation 21 (page 91)

1. B 2. G 3. D 4. G
5. B 6. Answers will vary.

Congruent Rectangles and Triangles (page 92)

1. a and b c and f
   d and e
2. a and e b and f
   c and d g and h

Quilting Bee (pages 93–94)

Problem A

Answers will vary. Check student work.

Problem B

Answers will vary. Check student work.

Problem C

Answers will vary. Check student work.

Answer Key *(cont.)*

Extension

Answers will vary. Check student work.

Standardized Test Preparation 22 (page 95)

1. C 2. F 3. C 4. G
5. C 6. Answers will vary.

Mirror Images (page 96)

Check answers for questions 1–5.

Challenge

Check student work for correct lines of symmetry.

1. A = 1 C = 1
   D = 1 E = 1
   H = 2 I = 2
   K = 1 M = 1
   O = infinite T = 1
   U = 1 V = 1
   W = 1 X = 2
   Y = 1
2. 0, 3, 8

Acting It Out or Using Concrete Materials Problems (page 97)

Problem A

6 pieces of paper

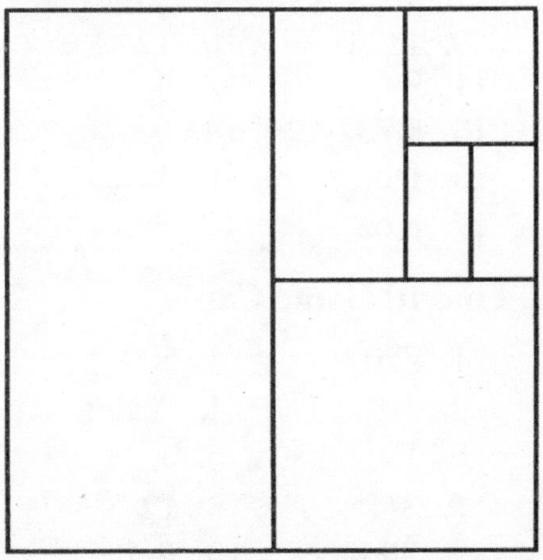

Problem B

one \$10 bill, one \$5 bill, four \$1 bills, and 4 quarters.

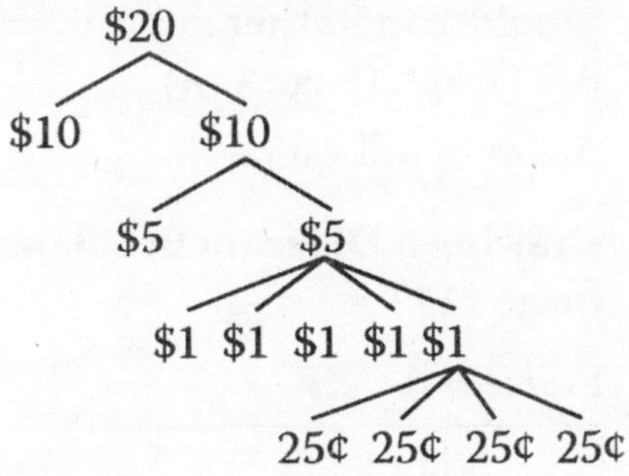

Standardized Test Preparation 23 (page 98)

1. D 2. G 3. C 4. G
5. B 6. Answers will vary.

Creating Patterns (page 99)

1. small circle
2. small square
3. small circle
4. large circle
5. square
6. sideways rectangle

Getting Around Town (pages 100–101)

Problem A

Answers will vary.; Sample Answer: Go east on Elm Street, go south on 3rd Avenue, and then go west on Main Street.

Problem B

Answers will vary.; Sample Answer: Go south on 1st Avenue and then go east on Main Street. Go east on Elm Street, south on 2nd Avenue, and east on Main Street.

Problem C

Answers will vary.; Sample Answer: Go east on Park Street, turn south on 3rd Avenue, and go west on Main Street. Continue west on Main Street, stop for gas, and then go north on 1st Avenue. Finally, go west on Elm Street to get home.

Standardized Test Preparation 24 (page 102)

1. D 2. H 3. D 4. F
5. C 6. Answers will vary.

Three-Dimensional Shapes (page 103)

Answers will vary depending on classroom discussion.

Acting It Out or Using Concrete Materials Group Problems (page 104)

Problem A

6 + 4 + 2 = 12; first plate = 6 cupcakes; second plate = 4 cupcakes; third plate = 2 cupcakes

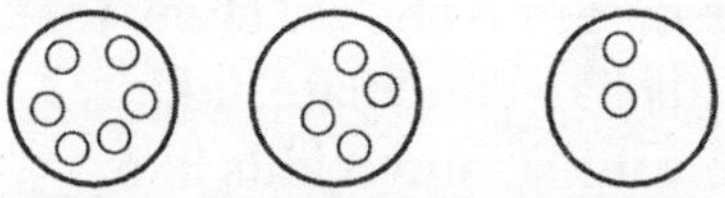

1st plate	2nd plate	3rd plate
$\frac{1}{2}$ of 12 = 6	6 – 2 = 4	(12 – 6) – 4 = 2

Standardized Test Preparation 25 (page 105)

1. D 2. J 3. D 4. G
5. C 6. Answers will vary.

Raffle Tickets (page 106)

1.

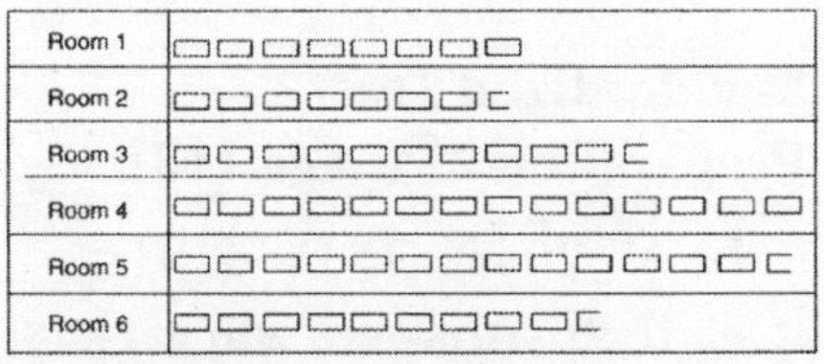

Answer Key *(cont.)*

2. a. Wednesday
   b. Thursday
   c. 280 tickets
   d. 3 more tickets

Drawing a Diagram (page 107)

Extension

8 rolls; 10 cuts on each roll or 20 cuts total

Standardized Test Preparation 26 (page 108)

1. C 2. H 3. A 4. G
5. D 6. Answers will vary.

Our Heights (page 109)

Consult classroom data for correct answers.

Graphing Height (page 110)

Answers will vary. Check students' graphs.

We All Scream for Ice Cream (pages 111–112)

Problem A

The diagram should show that 5 kids like chocolate, 4 kids like vanilla, and 3 kids like strawberry ice cream.

Problem B

chocolate is the favorite; cone is preferred

Problem C

3 kids

Extension

Answers depend on class data.

Standardized Test Preparation 27 (page 113)

1. A 2. H 3. D 4. H
5. B 6. Answers will vary.

What Would You Rather Do? (page 115)

Answers will vary.

Would You Rather . . . Bar Graph (page 116)

Answers will vary.

Drawing a Diagram Problems (page 117)

Problem A

	Jim		
			Billy
		Cheryl	
Robin			

Problem B

Check students' work; pizzas 2 & 3 are equal; pizzas 4 & 6 are equal

Standardized Test Preparation 28 (page 118)

1. C 2. H 3. A 4. H
5. B 6. Answers will vary.

Collecting Box Coupons (page 119)

1. 25 students
2. Dorothy
3. 750 box coupons
4. 2 5. 7 6. 50 CDs

Celebrating a Birthday (pages 120–121)

Problem A

$\frac{1}{2}$ of the cake is eaten

Problem B

2 pieces

Problem C

$\frac{1}{8}$ of the cake

Extension

Answer will depend on the number of students in your class.

Standardized Test Preparation 29 (page 122)

1. B 2. J 3. D 4. H
5. C 6. Answers will vary.

Will It or Won't It? (page 123)

1. The chance of something.
2. Answers will vary. Check students' work.

Drawing a Diagram Group Problems (page 124)

Problem A

Answers will vary.; Sample: a regular pentagon with 2 inch sides.

Standardized Test Preparation 30 (page 125)

1. A 2. F 3. D 4. H
5. D 6. Answers will vary.

Match It! Game Cards

1. 72 2. 5:00
3. 8 crackers
4. square pyramid
5. 930 6. $12.60
7. 4 x 5 = 20 8. 55
9. 11:20 10. 12
11. 9 12. 4
13. 5,732 14. 6
15. Dog 16. 10:15
17. 8 oz. 18. 8

Find It! Game Cards

1. 770 2. 500
3. 4/8 4. Water
5. 15 6. 3
7. 3/4 8. 12
9. Slide 10. Ben

Answer Key *(cont.)*

11. Figure 1	12. $19.25
13. 42	14. 10
15. 34	16. 5 x 4 = 20
17. 9 plants	18. 32 cm

Solve It! Game Cards

1. 379 cans
2. 7 cm
3. 30 cm
4. 50 students
5. 30 minutes
6. 37 cm
7. Oct. 16
8. Answers will vary. The hexagon could be smaller or larger than the given shape.
9. similar
10. subtract 100
11. 20 votes
12. 3
13. 1
14. reflection
15. 40 minutes
16. triangle
17. Answers will vary. The shaded parts of the rectangle should match the fraction written.
18. 14 oz.

Notes